Phase-Contrast and Dark-Field Imaging

Phase-Contrast and Dark-Field Imaging

Special Issue Editor

Simon Zabler

MDPI • Basel • Beijing • Wuhan • Barcelona • Belgrade

MDPI

Special Issue Editor
Simon Zabler
University of Würzburg
Germany

Editorial Office
MDPI
St. Alban-Anlage 66
Basel, Switzerland

This is a reprint of articles from the Special Issue published online in the open access journal *Journal of Imaging* (ISSN 2313-433X) in 2018 (available at: https://www.mdpi.com/journal/jimaging/special_issues/dark_field_imaging)

For citation purposes, cite each article independently as indicated on the article page online and as indicated below:

LastName, A.A.; LastName, B.B.; LastName, C.C. Article Title. *Journal Name* **Year**, *Article Number*, Page Range.

ISBN 978-3-03897-284-6 (Pbk)
ISBN 978-3-03897-285-3 (PDF)

Contents

About the Special Issue Editor

Simon Zabler studied Physics at Karlsruhe Germany and Grenoble France and holds a Double Diploma including a Master's degree from the Institut National Polytechnique de Grenoble, France. Zabler did his Doctorate at John Banhart's group in Berlin and worked as a postdoc at the Max Planck Institute for Colloids and Interfaces Potsdam, Germany. He has been teaching materials science at TU Berlin and joined the X-ray Microscopy labs LRM and Prof. Randolf Hanke in 2011 as assistant professor. Today, S. Zabler is managing the Fraunhofer group for NanoCT systems at Würzburg and is about to receive his habilitation.

Preface to "Phase-Contrast and Dark-Field Imaging"

Phase-contrast and dark-field imaging have been quickly-growing topics in the field of X-ray imaging for the past 15 years, both in medical and in materials communities. More than one hundred X-ray laboratories, and many synchrotron beamlines, are, today, equipped with Talbot-interferometers, Lau-interferometers, or Talbot-Lau setups. However, new setups emerge every day, e.g., based on the far-field moiré effect or on transmission X-ray microscopy. Both grating-based differential phase-contrast and dark-field image contrast have been shown to provide answers to formerly unsolvable questions in the field of medical and material imaging, e.g., imaging human lung tissue at the micrometer level or calculating local fiber orientation in carbon fiber composite automotive parts. Meanwhile, tomographic reconstruction and signal processing for these new contrast modes have become important fields of research in mathematics and computer science. Optimizing a grating-interferometer for a given task is not an easy task and requires detailed knowledge, both of the imaging physics and of the signal response from the specimen. Furthermore, the production processes for high-aspect optical gratings are presently at their technological limits, which will have to be overcome using entirely new approaches, if we want use DPC and DIC to inspect larger parts, e.g., from aircrafts and cars.

Simon Zabler
Special Issue Editor

Editorial

Phase-Contrast and Dark-Field Imaging

Simon Zabler

Fraunhofer Development Center X-ray Technology (EZRT) and Lehrstuhl für Röntgenmikroskopie (LRM), Josef-Martin Weg 63, 97084 Würzburg, Germany; simon.zabler@iis.fraunhofer.de

Received: 25 September 2018; Accepted: 25 September 2018; Published: 2 October 2018

Abstract: Very early, in 1896, Wilhelm Conrad Röntgen, the founding father of X-rays, attempted to measure diffraction and refraction by this new kind of radiation, in vain. Only 70 years later, these effects were measured by Ulrich Bonse and Michael Hart who used them to make full-field images of biological specimen, coining the term phase-contrast imaging. Yet, another 30 years passed until the Talbot effect was rediscovered for X-radiation, giving rise to a micrograting based interferometer, replacing the Bonse–Hart interferometer, which relied on a set of four Laue-crystals for beam splitting and interference. By merging the Lau-interferometer with this Talbot-interferometer, another ten years later, measuring X-ray refraction and X-ray scattering full-field and in cm-sized objects (as Röntgen had attempted 110 years earlier) became feasible in every X-ray laboratory around the world. Today, now that another twelve years have passed and we are approaching the 125th jubilee of Röntgen's discovery, neither Laue-crystals nor micrograting are a necessity for sensing refraction and scattering by X-rays. Cardboard, steel wool, and sandpaper are sufficient for extracting these contrasts from transmission images, using the latest image reconstruction algorithms. This advancement and the ever rising number of applications for phase-contrast and dark-field imaging prove to what degree our understanding of imaging physics as well as signal processing have advanced since the advent of X-ray physics, in particular during the past two decades. The discovery of the electron, as well as the development of electron imaging technology, has accompanied X-ray physics closely along its path, both modalities exploring the applications of new dark-field contrast mechanisms these days. Materials science, life science, archeology, non-destructive testing, and medicine are the key faculties which have already integrated these new imaging devices, using their contrast mechanisms in full. This special issue "Phase-Contrast and Dark-field Imaging" gives us a broad yet very to-the-point glimpse of research and development which are currently taking place in this very active field. We find reviews, applications reports, and methodological papers of very high quality from various groups, most of which operate X-ray scanners which comprise these new imaging modalities.

Keywords: X-ray phase-contrast imaging; X-ray scattering; dark-field imaging; Computed Tomography; Talbot-interferometer; coded-aperture imaging; Moiré pattern analysis; Electron Backscatter imaging; cultural heritage; medical imaging; image processing; fourier image analysis

1. An Introduction to This Special Issue

Following the Bonse-Hart X-ray interferometer, laboratory grating-based X-ray phase-contrast and dark-field imaging were first demonstrated in 2006 by Franz Pfeiffer's group at Paul–Scherrer Institute (PSI), Switzerland [1,2]. In fact, the dark-field modality was discovered two years later by the same group [3]. These developments were merely a translation of the previously introduced "Talbot-interferometer", invented by Atsushi Momose and first demonstrated by Christian David in 2002 [4,5]. It is worth noting that both Momose and David did their early experiments at the European Synchrotron Radiation Facility (ESRF). It was on the ID-19 beamline of this facility where the principal development and proof-of-concept of differential phase contrast (DPC) imaging based on Talbot-interferometry took place from 2002 to 2006. Twelve years later, thanks to the addition of a third (G0) grating making the Talbot-interferometer a Talbot–Lau interferometer, which overcomes the restrictions on source coherence, more than 100 academic laboratories reportedly host Talbot–Lau interferometers, performing experiments of this kind on a daily basis. Phase-contrast and dark-field imaging are about to become standard techniques.

1.1. Data, Methods and Results

One of the groups which exploits these new modalities is the Biomaterials Science Center at the University of Basel, Switzerland, where Bert Müller et al. employ micro computed tomography (CT) as well as grating-based phase-contrast (GBPC) for studying carious lesions in teeth, more precisely in the enamel [6]. Enamel is a nanocrystalline and prismatic arrangement of hydroxyapatite crystallites. This material features a highly anisotropic, oriented microstructure. Through the combination of micro-CT with scanning small-angle X-ray scattering (SAXS), the authors show how the lesions correlate (or not) with the orientation of the underlying structure. In addition to the very visual insights into carious lesions, their paper paves the ground for future applications (clinical or preclinical) of X-ray vector radiography (XVR), which is based on the anisotropy in dark-field image contrast (DIC). DIC has been shown to encode the same signal as ultra-small angle X-ray scattering (USAXS), which originates from micro- and nanometer-sized scatterers.

Veronika Ludwig from Gisela Anton's group (ecap) at Erlangen Friedrich Alexander University (FAU), on the other hand, presents the potential application of DIC and DPC to examine archeological findings comprising fabric and organic remnants [7]. This study is very valuable, first because it establishes an absolute scale of DIC for a whole array of fabrics and woven structures; and second because it addresses the optimization of the measurement protocol for performing DIC and DPC for different specimens. The group has been working on dose sensitive clinical applications such as mammography for long. Since X-ray inspections of archeological findings, in particular organic remnants, have been declared a dose sensitive issue (the inspection must not destroy DNA remnants) as well, Veronika Ludwig's study makes a perfect starting point for this application.

On the methodological side, we received surprisingly many contributions which dealt with "alternative acquisition schemes" to measure DPC and DIC, and only few on the "classical acquisition scheme" for Talbot–Lau imaging. One reason for this may be that the classical acquisition scheme includes three line gratings, each of which has to be positioned and oriented with respect to the X-ray beam axis as well as with respect to each other, hence 9 translational and 9 rotational degrees of freedom, some of which can be omitted, but others not, making X-ray GBPC a relatively bulky and expensive experiment. GBPC is further limited in terms of X-ray energy by the available aspect rations of the intensity gratings, which already reach 1:100 by current state-of-the-art LIGA (lithography and galvanic molding). An outstanding review on X-ray speckle-based imaging, written by Marie-Christine Zdora, a former student of Irene Zanette, from Diamond Light Source, is setting the ground for obtaining phase-contrast and dark-field data without using line gratings [8]. Speckle-based imaging is presented as a "single-shot" (XST) as well as a "stepping" (XSS) technique. These methods replace the grating with a simple diffusor, which may be sandpaper or a fibrous membrane. Unlike the conventional Talbot–Lau setup, XST and XSS yield two-dimensional DPC and DIC contrast.

The same methodological bifurcation, from stepping to single-shot, is reported by Maria Seifert from FAU. Seifert and her coauthors set the stage (meaning, the line gratings) to obtain very regular fine-spaced moiré fringes. The latter can also be analyzed from a single-shot, yielding DPC and DIC images at the expense of spatial resolution. While for XST this spatial resolution is linked to the speckle size, which has to be carefully adjusted, the improved image reconstruction by Seifert et al., from moiré patterns yields a relatively homogeneous resolving power. Distortions of the patterns caused by bent gratings are compensated for by their improved data processing chain [9].

Extending spackle-based imaging, in particular X-ray speckle tracking (XST), to higher energies in laboratory settings is the ambitious goal of Tunhe Zhou and Fei Yang, a close collaboration between the

Swiss EMPA (Eidgenössische Materialprüfungs- und Forschungsanstalt) and Diamond Light Source, UK [10]. Replacing the "standard" sandpaper which has been often used as diffusor for XST with steel wool, Zhou and her coauthors produce a random polychromatic speckle pattern in cone beam geometry with an average visibility of 17% at 80 kVp acceleration voltage, which is outstandingly good. The study shows applications of the technique to electronics (2D inspection) and to mortar (3D CT). These results are very impressive and extending their setup to XSS in addition to XST would indeed greatly improve spatial resolution.

Another nonconventional technique, namely Edge Illumination (EI), has evolved simultaneously to the Talbot–Lau interferometer, yielding images very similar to DIC and DPC. The report from Marco Endrizzi from Alessandro Olivio's group at University College London (UCL) is the latest hallmark in data analysis from EI imaging [11]. Endrizzi's method first reduces pixel sampling by binning to avoid artifacts which arise from high-frequency noise. Then, higher modulation amplitude and phase are progressively reconstructed by reducing the binning size until the desired pixel resolution is achieved. The result is a very stable reconstruction algorithm which also applies to single-shot EI image acquisition (here line scanning was used) and possibly to Talbot–Lau interferometers as well.

Jonas Dittmann from LRM (Lehrstuhl für Röntgenmikroskopie) Würzburg, Germany, presents a new algorithm which resolves a similar issue in Talbot-interferometry: The instrument, having very strong angular sensitivity with respect to the gratings positions, is commonly suffering from mechanical and thermal instabilities, either as a result of thermal drift or wobble of the phase-stepping axis [12]. Therefore, for example, during a CT scan, reference images are taken repeatedly and often. Dittmann uses the moiré patterns themselves to introduce a numerical self-alignment of the gratings positions and orientations. Thus, common artifacts, for example, residual moiré patterns, are eliminated and quality is improved. The algorithm is also well suited for a fine adjustment of the interferometer itself.

Lastly, Nicolas Brodusch from McGill University, Montréal Canada, guides us to a different use of the term dark-field imaging, thus widening the scope of this special issue beyond the domain of X-ray physics, namely to electron imaging. Brodusch's review on Electron Backscatter Diffraction (EBSD) and EBSD-DF (dark-field) imaging includes a very appealing explanation of how the BS electron diffraction patterns (EBSP) originate and are processed to yield a two-dimensional map of the lattice structure in polycrystalline materials. Just like the X-ray DIC, the darkfield contrast was only recently discovered in EBDS by selecting a small (dark) region in pseudo-Kikuchi diffraction patterns and plotting average intensity as a function of the electron spot coordinates. Today, EBSD-DF is gradually explored and recognized to encode very valuable physical information, for example, materials composition and/or surface topography (depending on the collection angle of the virtual aperture). Brodusch and his coauthors apply EBSD-DF to study minerals as well as magnetic domains in electrical steel, both in reflection. When EBSD-DF is applied in transmission (AA2099 Al-Li-Cu alloy), the visual effects which arising from this contrast mode are even more pronounced, revealing additional information about precipitates and deformation in the alloy's grain structure [13].

1.2. Quality and Impact

The eight research papers which were published in this special issue between March 2018 and August 2018, all with very short time spans between submission, reviews, and acceptance, feature a very high level of research and reporting quality. The manuscripts were reviewed quickly yet extensively by known experts in the field of X-ray imaging and electron imaging all throughout the world.

2. Further Reading

Readers interested in the topic of phase-contrast and darkfield imaging are referred to the book of David Paganin "Coherent X-ray optics" as well as to the pioneering publications by Atsushi Momose, Christian David, and Franz Pfeiffer [1–4,14]. For EBSD imaging, one might refer to the book by Schwartz et al. [15].

Funding: This research received no external funding.

Acknowledgments: The guest editor thanks all contributing authors for dedicating their work and time to this special issue as well as the anonymous peer reviewers for their prompt yet elaborate remarks, who were keeping the presentation standard for this issue constantly high. Assistant Editor Veronica Wang, a lot of thanks for her tireless efforts and communications throughout the editing process.

Conflicts of Interest: The author declares no conflict of interest.

References

1. Bonse, U.; Hart, M. An X-ray interferometer. *Appl. Phys. Lett.* **1965**, *6*, 155–156. [CrossRef]
2. Pfeiffer, F.; Weitkamp, T.; Bunk, O.; David, C. Phase retrieval and differential phase-contrast imaging with low-brilliance X-ray sources. *Nat. Phys.* **2006**, *2*, 258–261. [CrossRef]
3. Pfeiffer, F.; Bech, M.; Bunk, O.; Kraft, P.; Eikenberry, E.F.; Brönnimann, C.; Grünzweig, C.; David, C. Hard-X-ray dark-field imaging using a grating interferometer. *Nat. Mater.* **2008**, *7*, 134–137. [CrossRef] [PubMed]
4. Momose, A.; Kawamoto, S.; Koyama, I.; Hamaishi, Y.; Takai, K.; Suzuki, Y. Demonstration of X-ray Talbot interferometry. *Jpn. J. Appl. Phys.* **2003**, *42*, L866–L868. [CrossRef]
5. David, C.; Nöhammer, B.; Solak, H.; Ziegler, E. Differential X-ray phase contrast imaging using a shearing interferometer. *Appl. Phys. Lett.* **2002**, *81*, 3287–3289. [CrossRef]
6. Deyhle, H.; White, S.; Botta, L.; Liebi, M.; Guizar-Sicairos, M.; Bunk, O.; Müller, B. Automated Analysis of Spatially Resolved X-ray Scattering and Micro Computed Tomography of Artificial and Natural Enamel Carious Lesions. *J. Imaging* **2018**, *4*, 81. [CrossRef]
7. Ludwig, V.; Seifert, M.; Niepold, T.; Pelzer, G.; Rieger, J.; Ziegler, J.; Michel, T.; Anton, G. Non-Destructive Testing of Archaeological Findings by Grating-Based X-Ray Phase-Contrast and Dark-Field Imaging. *J. Imaging* **2018**, *4*, 58. [CrossRef]
8. Zdora, M.C. State of the Art of X-ray Speckle-Based Phase-Contrast and Dark-Field Imaging. *J. Imaging* **2018**, *4*, 60. [CrossRef]
9. Seifert, M.; Gallersdörfer, M.; Ludwig, V.; Schuster, M.; Horn, F.; Pelzer, G.; Michel, T.; Anton, G. Improved Reconstruction Technique for Moiré Imaging Using an X-Ray Phase-Contrast Talbot–Lau Interferometer. *J. Imaging* **2018**, *4*, 62. [CrossRef]
10. Zhou, T.; Yang, F.; Kaufmann, R.; Wang, H. Applications of Laboratory-Based Phase-Contrast Imaging Using Speckle Tracking Technique towards High Energy X-Rays. *J. Imaging* **2018**, *4*, 69. [CrossRef]
11. Endrizzi, M.; Vittoria, F.; Olivo, A. Single-Shot X-ray Phase Retrieval through Hierarchical Data Analysis and a Multi-Aperture Analyser. *J. Imaging* **2018**, *4*, 76. [CrossRef]
12. Dittmann, J.; Balles, A.; Zabler, S. Optimization based evaluation of grating interferometric phase stepping series and analysis of mechanical setup instabilities. *J. Imaging* **2018**, *4*, 77. [CrossRef]
13. Brodusch, N.; Demers, H.; Gauvin, R. Imaging with a Commercial Electron Backscatter Diffraction (EBSD) Camera in a Scanning Electron Microscope: A Review. *J. Imaging* **2018**, *4*, 88. [CrossRef]
14. Paganin, D. *Coherent X-ray Optics*; Oxford Science Publications: New York, NY, USA, 2006.
15. Schwartz, A.J.; Kumar, M.; Adams, B.L.; Field, D.P. *Electron Backscatter Diffraction in Materials Science*; Springer: New York, NY, USA, 2009.

Journal of
Imaging

MDPI

Article

Automated Analysis of Spatially Resolved X-ray Scattering and Micro Computed Tomography of Artificial and Natural Enamel Carious Lesions

Hans Deyhle [1], Shane N. White [2], Lea Botta [1], Marianne Liebi [3], Manuel Guizar-Sicairos [3], Oliver Bunk [3] and Bert Müller [1,*]

[1]   Biomaterials Science Center, University of Basel, Gewerbestrasse 14, 4123 Allschwil, Switzerland;
      hans.deyhle@unibas.ch (H.D.); Lea.Botta@uzb.ch (L.B.)
[2]   UCLA School of Dentistry, University of California, 10833 Le Conte Ave., Los Angeles, CA 90095-1668, USA;
      snwhite@dentistry.ucla.edu
[3]   Paul Scherrer Institute (PSI), 5232 Villigen, Switzerland; marianne.liebi@chalmers.se (M.L.);
      manuel.guizar-sicairos@psi.ch (M.G.-S.); oliver.bunk@psi.ch (O.B.)
*   Correspondence: bert.mueller@unibas.ch; Tel.: +41-61-207-5431

Received: 10 April 2018; Accepted: 13 June 2018; Published: 15 June 2018

Abstract: Radiography has long been the standard approach to characterize carious lesions. Spatially resolved X-ray diffraction, specifically small-angle X-ray scattering (SAXS), has recently been applied to caries research. The aims of this combined SAXS and micro computed tomography (µCT) study were to locally characterize and compare the micro- and nanostructures of one natural carious lesion and of one artificially induced enamel lesion; and demonstrate the feasibility of an automated approach to combined SAXS and µCT data in segmenting affected and unaffected enamel. Enamel, demineralized by natural or artificial caries, exhibits a significantly reduced X-ray attenuation compared to sound enamel and gives rise to a drastically increased small-angle scattering signal associated with the presence of nanometer-size pores. In addition, X-ray scattering allows the assessment of the overall orientation and the degree of anisotropy of the nanostructures present. Subsequent to the characterization with µCT, specimens were analyzed using synchrotron radiation-based SAXS in transmission raster mode. The bivariate histogram plot of the projected data combined the local scattering signal intensity with the related X-ray attenuation from µCT measurements. These histograms permitted the segmentation of anatomical features, including the lesions, with micrometer precision. The natural and artificial lesions showed comparable features, but they also exhibited size and shape differences. The clear identification of the affected regions and the characterization of their nanostructure allow the artificially induced lesions to be verified against selected natural carious lesions, offering the potential to optimize artificial demineralization protocols. Analysis of joint SAXS and µCT histograms objectively segmented sound and affected enamel.

Keywords: enamel caries; small-angle X-ray scattering; image registration; bivariate histogram plot; segmentation; multi-modal imaging

1. Introduction

Tooth enamel, a unique body tissue, presents some distinctive challenges to study. Compared to other human tissues, it is extremely dense and homogenous, comprising almost entirely of elongated hydroxyapatite crystallites. The organization of tooth enamel is particularly complex with orientation and structure at nanometer, micrometer, and millimeter levels, but the remarkably uniform composition obscures structural subtlety to most forms of examination.

Carious dissolution of tooth enamel, the most common disease to afflict mankind, has been studied since the late 19th century [1–3]. Caries detection, characterization and diagnosis remain a problematical issue [4]. Diagnosis is the art of identifying a disease through signs and symptoms, but early enamel caries may present few if any symptoms to the patient and few if any signs to the clinician. Hence, much attention has focused upon caries detection, primarily through radiography and optical inspection. In a clinical setting, radiographic appearance alone, specifically the depth of radiolucency, is often used to make a decision as whether to treat or not.

Radiographic sensitivity to early stage subsurface lesions, however, is limited and often even inadequate [5]. Substantial carious dissolution must occur before the lesion is reliably detected in vivo [6]. Attenuation coefficient in projection might therefore be insufficient. Scattering arising from micro/nano-porosity provides a different type of contrast. It is reasonable to hypothesize that the combination of X-ray attenuation and scattering signals allows for a better caries detection.

For in vitro detection, more sensitive X-ray methods can be used, because the X-ray dose is hardly relevant, thus allowing the exploration of alternative methods for caries identification, still a matter of investigation [7–9]. Improved objective measures of the early carious lesion would be of inestimable clinical and research utility [10].

Spatially resolved micro-beam small-angle (SAXS) and wide-angle (WAXS) X-ray scattering was first applied to complex hierarchically organized biological structures two decades ago [11]. Such reciprocal-space techniques have been frequently used to analyze calcified tissues ex vivo, often combined with more or less surface sensitive electron and light microscopies. SAXS and WAXS yield information complementary to hard X-ray transmission (radiography) [12–14] and similar to that obtained in grating-based X-ray dark-field imaging [15].

Likewise, micro-beam diffraction techniques have been applied to nanostructural and crystallographic investigations of healthy enamel as well as on artificially induced and natural caries [16–22]. The degree of co-alignment of hydroxyapatite crystallites within unaffected and carious enamel has been quantified using WAXS [17,19]. Crystallite loss, measured using WAXS, has been related to void formation, measured using SAXS, in subsurface lesions [22]. In such studies, information from complementary techniques was compared, but classification of enamel as being either carious or unaffected was performed using one selected standard alone. We propose combining X-ray imaging, i.e., radiography and micro computed tomography (μCT) with spatially resolved SAXS to segment the carious enamel on about 0.5 mm-thick crown slices. Here, we use radiographic projections obtained by forward-projecting μCT data to enhance the segmentation of the two-dimensional SAXS data. The three-dimensional data is ultimately not necessary, but useful for validation purposes. Data from spatially resolved SAXS and μCT have not previously been combined with a (semi-)automated histogram analysis approach to segment affected and unaffected tissues. Both techniques have previously been used independently for the in vitro characterization of carious lesions [23–26].

Micro-beam diffraction was proposed as an analytic tool, e.g., in bone [27], or for breast [28] and brain cancer [29] characterization. Our approach could extend these methods for the segmentation of normal and diseased, or normal and repaired tissues.

Hence, we comparatively study one artificial lesion prepared within days and one selected natural carious lesion formed over a comparatively much longer period of time in the order of months or even years by means of the complementary experimental techniques SAXS and radiography/μCT using specifically developed software for their combined analysis. The goal of the investigation is to demonstrate the feasibility of the (semi-)automated analysis of the joint SAXS and forward-projected μCT histogram to localize normal and affected enamel within a crown slice of clinically relevant thickness. This approach also extends the scope of application for both characterization techniques to an artificially induced lesion and the additional benefit resulting from dedicated data analysis of their combination. The questions of whether and how far quickly generated artificial lesions correspond to natural lesions is, thus, directly addressed for the first time using the combination of multiple parameters.

2. Materials and Methods

2.1. Ethical Approval

A naturally occurring surface carious lesion, a white spot, and an artificially induced carious lesion, obtained from second and third molars, were studied. All procedures were conducted in accordance with the Declaration of Helsinki and according to the ethical guidelines of the Canton of Basel, Switzerland. The responsible Ethical Committee approved the study with the number 290/13. The teeth were previously scheduled for extraction for clinical reasons unrelated to this study. Patients gave written consent for the use of their extracted teeth in the registration form of the Volkszahnklinik in Basel, Switzerland. The donated teeth were anonymized.

2.2. Sample Preparation

Immediately after extraction, the teeth were immersed in a 0.1% thymol solution. Soft tissue, calculus, and alveolar bone remaining on the extracted teeth were removed using a scalpel. The artificial lesion was defined by painting the tooth with a layer of nail varnish, leaving a window about 2 mm × 2 mm in size, see reference [29]. Subsequently, the tooth was incubated for three days in an acidic demineralization buffer (50 mM acetic acid, 2.2 mM $CaCl_2$, 2.2 mM NaH_2PO_4, titrated with 1 M KOH to pH 4.4) [30]. All chemicals were supplied by Sigma-Aldrich Co. LLC (Sigma-Aldrich Chemie GmbH, Buchs, Switzerland). Tooth slices were cut using a band saw (Exakt Apparatebau GmbH, Norderstedt, Germany). The slices, 1300 and 500 μm thin, were stored in water before and during measurements to prevent drying. The schemes in Figure 1a,b show the sample preparation steps for the teeth with artificial and natural lesion, respectively.

2.3. Micro Computed Tomography

The tooth slices were individually transferred into a deionized water-filled Eppendorf tube to maintain a wet environment and prevent drying. This Eppendorf tube was glued onto the holder of the manipulator in the μCT-system. Micro computed tomography data sets were acquired using a nanotom® m (phoenix | X-ray, GE Sensing and Inspection Technologies GmbH, Wunstorf, Germany) [31]. The voxel length corresponded to 7.0 μm. A 0.2 mm-thick copper filter was placed into the beam path to increase the mean photon energy and reduce beam hardening. For all specimens, the acceleration voltage was set to 90 kV with a tungsten-on-diamond target.

The acquired data were reconstructed using phoenix datos | x 2.0 reconstruction software (phoenix | X-ray, GE Sensing & Inspection Technologies GmbH, Wunstorf, Germany). Segmentation through thresholding was performed in MATLAB (2014a, MathWorks, Natick, MA, USA) in order to extract information about the magnitude of the local attenuation coefficients, directly related to density, within the selected parts of the teeth. The performance of the advanced laboratory μCT system for the analysis of crown of human teeth is comparable to synchrotron radiation-based setups [32].

2.4. Small-Angle X-ray Scattering Data Acquisition

Spatially resolved small-angle X-ray scattering measurements (SAXS) were performed at the cSAXS beamline of the Swiss Light Source (Paul Scherrer Institute, Villigen, Switzerland) [33]. The specimens were stored in polyimide sachets to keep the specimens hydrated, and raster-scanned in 30 μm × 10 μm steps in *x*- and *y*-directions (cf. Figure 1) through a monochromatic X-ray beam, with a photon energy of 18.6 keV, focused to 30 μm × 10 μm full-width-at-half-maximum spot size at the specimen location. The specimen to detector distance D_{sd} of 7.1 m (cf. Figure 1c) was determined with the first-order scattering ring of silver-behenate powder. With this setup, *q*-ranges corresponding to real-space periodicities (*d*-spacing) from 4 to 180 nm were investigated. To reduce the air scattering, an evacuated flight tube was placed between specimen and detector. A diode on the beam stop in front of the detector recorded the transmitted intensity of the X-ray beam during data acquisition. SAXS data treatment was performed with the cSAXS Matlab package available at https://www.psi.ch/sls/csaxs/software [33]. Additionally, the degree of orientation of the point-symmetric scattering patterns was defined as $1 - (\text{FWHM}_{\text{deg}}/180°)$, where FWHM_{deg} denotes the full-width-at-half maximum (FWHM) of the azimuthal SAXS intensity [17].

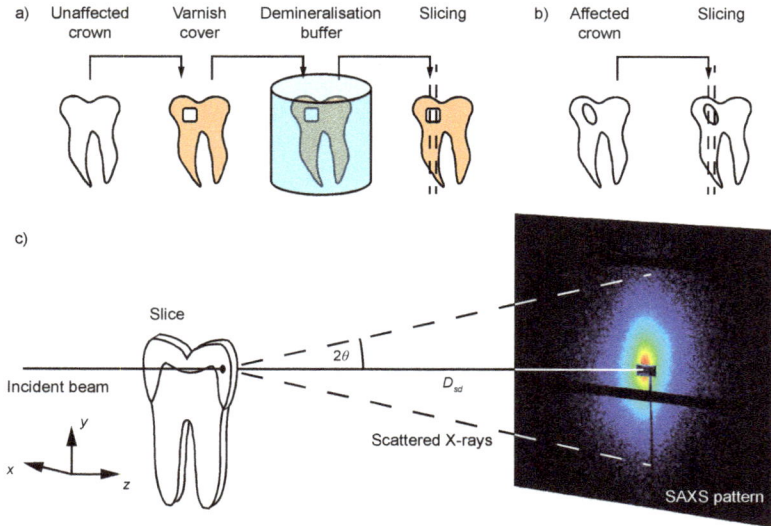

Figure 1. Sample preparation steps to obtain tooth slices with artificial (**a**) and natural (**b**) lesions and schematic representation of the spatially resolved SAXS set-up (**c**). The tooth slice is raster-scanned through the focused X-ray beam. At each position a two-dimensional scattering pattern is acquired. The direct beam transmitted by the sample is measured in intensity and absorbed by the beam stop in front of the detector.

2.5. Segmentation

The carious regions were segmented from the volumetric μCT data of the tooth slices via thresholding. To increase the contrast-to-noise ratio, a $5 \times 5 \times 5$ median filter was applied to the data prior to thresholding.

In addition, bivariate histogram plots of the total scattered intensity from the SAXS measurements and the X-ray attenuation within the tooth slices were generated. For this purpose, the local X-ray attenuation values of the reconstructed μCT data of the slices were integrated along the direction perpendicular to the slice (z-direction, cf. Figure 1). The bivariate histogram plots were segmented using the *k*-means clustering algorithm [34] implemented in the MATLAB statistics and machine learning toolbox. For this purpose, scattering intensities and X-ray attenuation data were rescaled to an arbitrary scale so that they presented the same minimum and maximum values, i.e., 1 and 200. This bin size was chosen to allow for a reasonable representation of the data, allowing the distinction of clusters without compromising segmentability through excessive noise. The *k*-means algorithm assigned each count in the bivariate histogram plot to one of three clusters. For each cluster, we defined an ellipse aligned along the eigenvectors of the covariance matrix and the length of the ellipse axes by three times the square root of the eigenvalues. The three ellipses were associated with the unaffected enamel, the dentin, and the lesion.

3. Results

Selected slices through the three-dimensional data from μCT, obtained from crowns with a natural and an artificial lesion, are illustrated in Figure 2a,b, respectively. The slices show the microstructure of lesions owing to the reduced mineral content. The mineralized, micrometer-thick outer shell encloses the body of the lesion. The natural lesion is about 250 μm deep, whereas the artificial lesion only extends about 50 μm.

Attenuation histograms of the entire μCT-datasets from the crown specimens with the natural and the artificial caries lesions are shown in the diagrams of Figure 2c,d, respectively. Enamel and dentin can be clearly discriminated; both exhibit the characteristic Gaussian distribution [35]. The enamel lesions exhibit attenuation values between those of dentin and enamel.

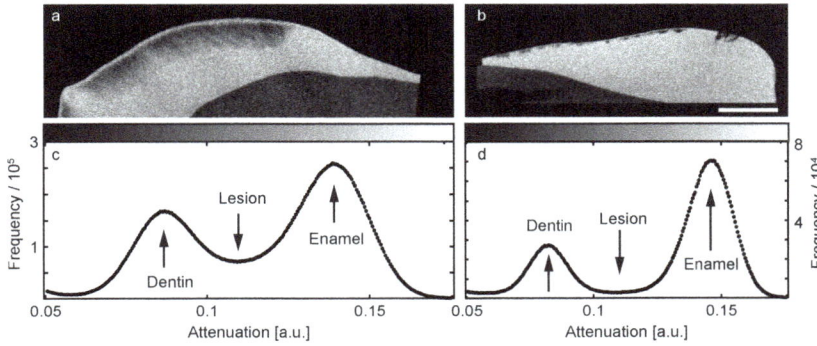

Figure 2. Selected slices through the μCT datasets of tooth slices with natural (**a**) and artificial (**b**) lesions. Affected regions exhibit X-ray attenuation reduced with respect to the healthy enamel. The related histograms of the three-dimensional data are shown in the diagrams (**c,d**). The length of the bar corresponds to 1 mm.

Subsequent to the characterization using μCT, the tooth slices were investigated by means of synchrotron radiation-based spatially resolved small-angle X-ray scattering in transmission mode to evaluate the anisotropy and the orientation of the nanostructures, foremost hydroxyapatite crystallites, within carious and unaffected enamel. Figure 3 shows the integrated scattered intensity at selected q-ranges corresponding to the real-space periodicities between 10 and 20 nm (a) and (d), 70 and 80 nm (b) and (e), and 140 and 150 nm (c) and (f). The images contain distinctive anatomical features, which allow correlating both datasets, as they are similar to those of the μCT-data. Both the natural and the artificially induced lesions exhibit an increased scattering intensity. These bright regions of higher scattering intensity correspond to the less mineralized and therefore darker regions in the μCT data (cf. Figure 2).

The main orientation of the scattering signal at each pixel position is displayed in Figure 4, according to the color-wheel. Since the scattering signal is related to the Fourier transform of the electron density of the specimen, the scattering signal orientation is perpendicular to the long extension of the nanoscale scatterers in the plane perpendicular to the X-ray beam, cf. x-y-plane in Figure 1c. The nanostructures in the enamel have an orientation that is radial to the dentin-enamel junction. The lesions that were prominently seen in the scattering intensity plots, cf. Figure 3, cannot be discerned from sound enamel in the orientation plots, see Figure 4. This indicates that the main orientation of the scatterers is preserved in both lesions.

The degree of anisotropy describes the co-alignment of the nanoscale structures, on average, over the area illuminated by the X rays, i.e., over several square-micrometers. A band of low anisotropy at the location of the dentin–enamel junction (DEJ) clearly delineates dentin and enamel. At larger scattering angles, corresponding to 10–20 nm real-space periodicities, the natural lesion and the sound enamel show similar values of degree of orientation, as represented in Figure 5a. For smaller scattering angles, i.e., larger periodicities, the differences between unaffected and carious enamel become distinct, see Figure 5b,c for 70–80 nm and 140–150 nm. The degree of anisotropy of carious enamel is around

0.7 for the *q*-ranges investigated, whereas the one of sound enamel decreases from 0.7 to 0.6 with decreasing *q*. In addition, one finds bands of alternating degree of orientation extend from the DEJ to the tooth surface, associated with the Hunter–Schreger bands. The artificial lesion is hardly identifiable at some investigated *q*-ranges, presenting regions of higher and lower anisotropy compared to the unaffected enamel, Figure 5d–f.

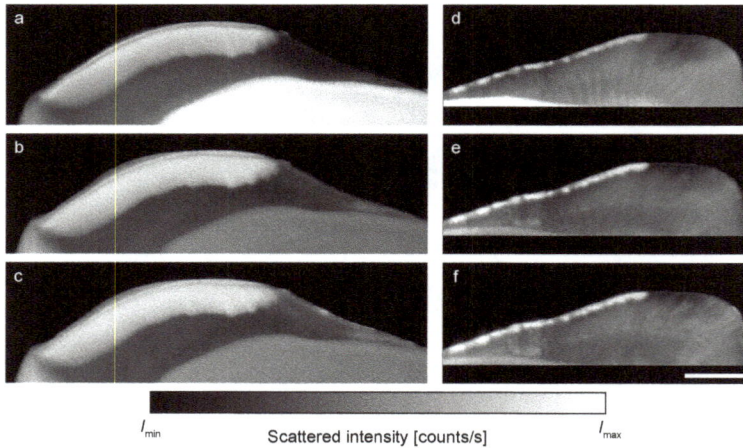

Figure 3. Integrated scattered intensity for the *q*-ranges that corresponds to the real-space periodicities between 10 and 20 nm (**a,d**), 70 and 80 nm (**b,e**), and 140 and 150 nm (**c,f**) for the natural (**left column**) and the artificial (**right column**) lesions. For all ranges, the lesion appears as bright band in the upper part of the image, corresponding to high scattering intensity. The enamel yields lower scattering intensities compared to dentin and lesion. The intensities are logarithmically scaled with I_{min} and I_{max} corresponding to (**a**) 2.0–53.3, (**b**) 0.2–10.0, (**c**) 33.4–26,500.0, (**d**) 8.4–2,650.0, (**e**) 66.7–133,000.0, and (**f**) 21.1–13,000.0 counts per pixel, respectively. The length bar corresponds to 1 mm.

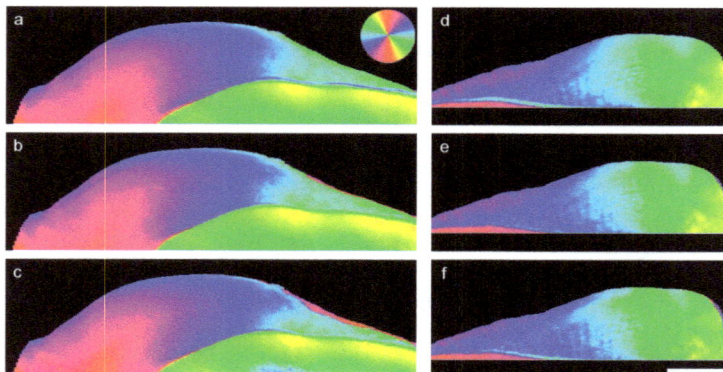

Figure 4. The color-coded images show the main azimuthal orientation of the scattering patterns for the *q*-ranges corresponding to the real-space periodicities between 10 and 20 nm (**a,d**), 70 and 80 nm (**b,e**), and 140 and 150 nm (**c,f**) for the natural (**left column**) and the artificial (**right column**) lesions. The direction of the scattered intensity is according to the color wheel in the inset. The carious lesions cannot readily be distinguished from the sound enamel, suggesting that the predominant orientation of the features on the investigated nanometer range is unaffected by the carious lesions generated. The scale bar corresponds to 1 mm.

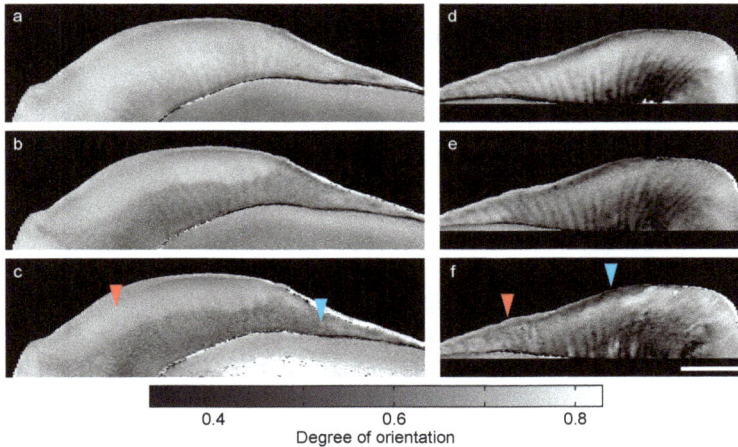

Figure 5. The gray-scale images display the degree of orientation of the scattering patterns for the *q*-ranges corresponding to the real-space periodicities between 10 and 20 nm (**a,d**), 70 and 80 nm (**b,e**), and 140 and 150 nm (**c,f**) for the natural (**left column**) and the artificial (**right column**) lesions. For the natural lesion the degree of orientation at lower scattering angles, images (**b,c**), is increased (red-colored arrow) compared to the sound enamel (blue-colored arrow), whereas for the artificial lesion regions with higher (red-colored arrow) and lower (blue-colored arrow) anisotropy compared to the sound enamel appear for all length scales investigated. The scale bar corresponds to 1 mm.

The local attenuation values integrated along the beam for the natural (a) and artificial (d) lesions and the related spatially resolved X-ray scattering intensities are shown in Figure 6b,e, respectively. The natural lesion appears as a dark region in the image of Figure 6a and bright in the image of Figure 6b. The artificially induced demineralization results in strongly enhanced SAXS intensity, see image in Figure 6e, (red-colored arrow). However, the artificial lesion was barely visible in the X-ray attenuation signal, represented in the image of Figure 6d. Only some spotted regions with reduced attenuation are present near the enamel surface of the artificial lesion. The images in Figure 6c,f show the thickness of the specimens' unaffected hard tissues in yellow color and the thickness of the lesions in blue color as obtained from the volumetric μCT data. The image in Figure 6f reveals several regions of reduced thickness within the enamel, which present reduced attenuation (blue-colored arrow). The thickness of the hard tissue histological section or slices from the crown directly affects the scattered X-ray intensity and the X-ray attenuation. In contrast to the scattering signal, without the additional 3D information from μCT, it is impossible to determine whether the reduced attenuation within the enamel is caused by demineralization or by variations in the slice thickness.

The bivariate histogram plots for the two slices of the crowns are represented in Figure 7. First, the X-ray attenuation obtained from the μCT measurements was integrated along the beam direction (z-direction, cf. Figure 1), yielding a two-dimensional image akin to radiography. Then, it was plotted for each pixel position against the related scattered intensities, where the *q*-range corresponds to real-space periodicities between 5 and 150 nm. The frequency of counts is displayed by the brightness (see gray-scale bars). Since both attenuation and SAXS intensity are dependent on specimen thickness, scaling of the abscissa and ordinate are adapted accordingly. For the natural lesion, cf. Figure 7a, one can easily discriminate between the three clusters. For the artificially induced lesion such discrimination is more complex, cf. Figure 7b. Nevertheless, the bivariate histogram plots offer a better approach for segmenting the lesions than the individual histograms given by integration along abscissae or ordinates; see the related diagrams at the borders of Figure 7.

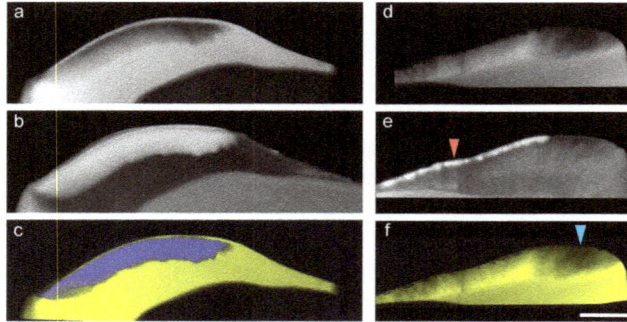

Figure 6. Images (**a**,**d**) show the lateral μCT attenuation values integrated across the slice thickness. The mean scattered intensities for the natural (**b**) and the artificial lesions (**e**) for the *q*-range correspond to the real-space periodicities between 5 and 150 nm. The thicknesses of the lesion in blue color and of sound enamel in yellow color are given in (**c**,**f**). Brighter color indicates larger thickness, being black for zero and bright yellow/blue for 1330 μm (**c**) and 413 μm (**f**), respectively. Mixed colors indicate regions where unaffected enamel and caries are present along the X-ray beam (z-direction, cf. Figure 1). The scale bar corresponds to 1 mm.

Next, each bivariate histogram plot was divided into three regions using the *k*-means algorithm implemented in MATLAB code. For each of the three regions the center of gravity *C* and moments of inertia were calculated. They define ellipses centered in *C* with the main axes in the direction of the moments of inertia and a magnitude three times the square root of the eigenvalues of the covariance matrix. All points within each ellipse were assigned to one region within the selected piece of the crown. The results are displayed in Figure 8. The red color marks the enamel affected by caries. The blue color relates to the unaffected enamel. For the crown slice containing the natural lesion, the yellow color corresponds to the dentin and to the intact surface layer. For the specimen containing the artificial lesion, cyan is associated with the enamel regions of reduced thickness. In the slice containing the artificial lesion, dentin is not correctly identified. Rather, thin enamel is identified as separate component. Nevertheless, a distinction is made between caries-affected and unaffected enamel.

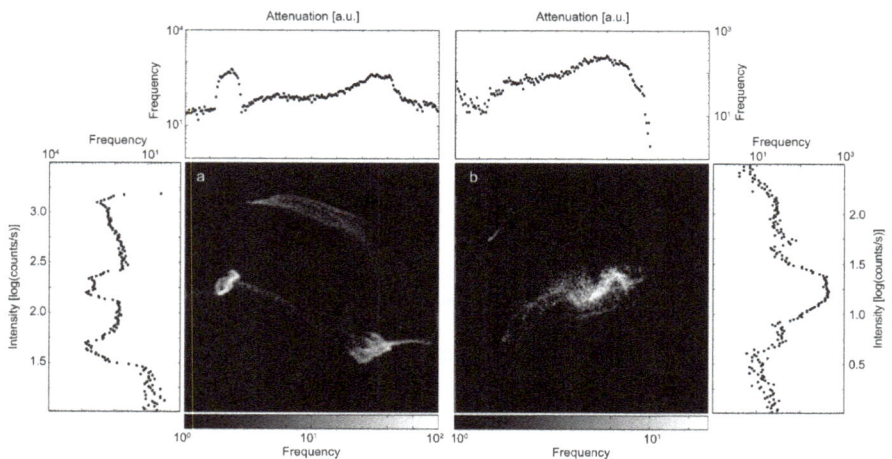

Figure 7. The bivariate histogram plots of the scattering signal (ordinate) and the integrated X-ray attenuation (abscissa) are displayed for the natural (**a**) and the artificial (**b**) lesions.

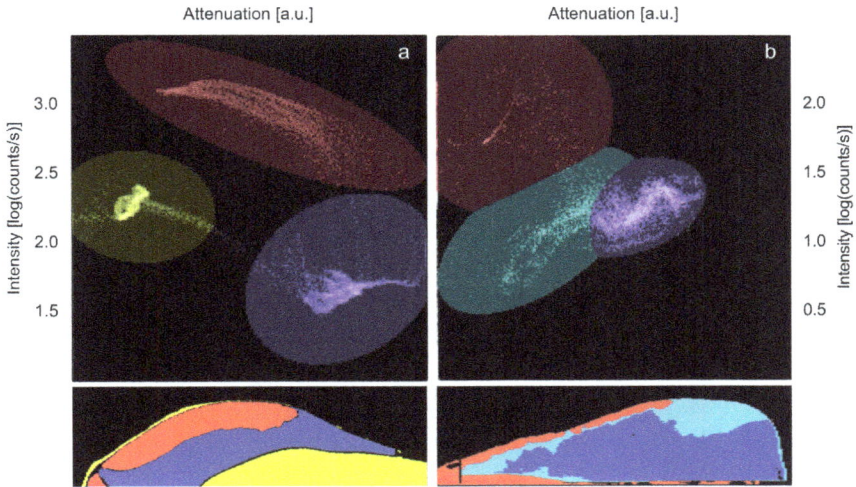

Figure 8. The bivariate histogram plots allow segmenting some anatomical features and identifying regions that belong to the carious and the unaffected enamel. (**a**) The red color indicates the lesions, blue color is associated with the unaffected enamel, and the yellow color corresponds to dentin. In the slice containing the artificial lesion (**b**), dentin is not correctly identified. Rather, it is associated with the lesions, whereas the thinned enamel is assigned to an independent component.

4. Discussion

Natural and appropriately prepared artificial carious lesions exhibit a decreased density below intact surfaces, which gives rise to a decrease in X-ray attenuation because of the reduced mineral density. It is frequently observed that the X-ray scattering potential of enamel lesions is significantly higher than that of sound enamel, associated with the increased nanometer-scale porosity within the lesions [17,20,21,36]. It has been a matter of debate whether the SAXS signal originates from the crystallites or the voids in between them. Several studies report that at larger q, the scattering signal is often associated with the voids, since the crystallites are large compared to the corresponding ranges [15,20,21,37]. Yagi et al. also stated that increased SAXS signal depends more on the surface area than on void volume [21]. The setup in this study allowed us to access smaller q-values, down to 0.04 nm^{-1}, corresponding to about 150 nm, well above the putative pore size of 5 nm [15]. We believe that the observed increase in anisotropy towards larger periodicities (cf. Figure 5) is associated with preferential etching of more arbitrarily oriented calcium phosphates within enamel between the rods. There is, however, consensus that the voids between rods are aligned to the rods themselves [20]. Our results support this notion, as we do not observe any change in the orientation of the anisotropic, nanometer-size components between the healthy and the diseased part of the crown, implying that the pores have the same orientation as the mineral crystallites. However, the pores do not have the same size and shape as the crystallites, because the experimental values for the degree of orientation differ between carious and healthy enamel.

Projections, as generated radiographically, do not allow for the discrimination between thickness-dependent and density-dependent phenomena. In Figure 6b, for example, the prominent dark region on the top left of the specimen, caused by reduced specimen thickness, cannot be distinguished from caries-affected regions if the geometry of the slice is unknown. Also, overlapping carious and unaffected enamel might lead to a misinterpretation or underestimation of the local degree of demineralization. Using the µCT data, the thickness is derived with micrometer precision, and the projected densities of overlapping features can be separated.

The segmentation of anatomical features with comparable X-ray attenuation and X-ray scattering potential is challenging. Especially the boundaries between features with gradual density changes, such as those between carious lesions and sound enamel, are difficult to identify. The exact registration of data from different modalities [38,39] or the data from the same modality obtained before and after tissue alteration permits the creation of a bivariate histogram plot for powerful feature segmentation [40]. Here, we have demonstrated that through the combination of X-ray attenuation and SAXS signal over a large q-range one can determine the size and shape of the natural and caries lesions in a (semi-)automated way, not biased by subjective interpretation of experts. This approach could be applied to other biological tissues, such as bone.

In the natural lesion, the presented approach reproduces the intact surface lesion, which is seen in the CT and radiography data, but is not reproduced in the SAXS intensity signal. Conversely, the region presenting reduced X-ray attenuation in the specimen with the artificial lesion was clearly identified as unaffected enamel by the segmentation process. Note that this approach assumed three regions for each specimen, namely sound dentin, sound enamel, and enamel lesion. While it succeeded in discriminating these regions in the specimen containing the natural lesion (cf. Figure 8a), it failed to do so for the specimen with the artificial lesion (Figure 8b). Here, strong variations in specimen thickness led to the dentin being associated with the affected enamel. Nevertheless, carious and sound enamel could be distinguished.

The natural and the artificial lesions of the present study show comparable characteristics, i.e., intact surface layers, reduced X-ray attenuation and increased X-ray scattering intensity within the lesions, but they also exhibit differences in size and shape. Such differences are even found for in vitro tests with artificial lesions and are manifested by inter-lab variability [41,42], implying that the challenging biomimetic in vitro generation of carious lesions. Thus, in situ reproduction of lesions was proposed [43]. Furthermore, the classically described lesion is by no means representative of all observed natural lesions [44]. This means that both in vitro and clinical studies should include a rather large and varied number of subjects before generalizable evidence is reached.

5. Conclusions

The present communication is not only a simple comparison of one artificially induced and one natural carious lesion but also describes a methodological approach combining radiographic data, spatially resolved X-ray scattering data and data treatment for the quantification and segmentation of lesions. The increased porosity within the lesion leads to a reduced X-ray attenuation and increased scattering intensity over the investigated q-ranges. The combination of the two results gives rise to a better differentiation within the projection data. This approach could be applied to other tissues.

Although a position-resolved SAXS setup, as presented in this study, is not well suited for a clinical environment, the future potential of the SAXS signal for caries diagnosis must not be underestimated. Other setup types, which are sensitive to small-angle scattering, as for example the dark-field signal obtained in X-ray Talbot–Lau interferometry measurements [45], might be more easily adapted for the in vivo diagnosis of mineral loss. The dark-field signal originates from scattering from entities smaller than the detector's pixel size [46]. SAXS signal integrated over a large q-range is similar and related to the dark-field signal obtained from X-ray Talbot–Lau interferometry [47], and thus, similar results can be expected.

Author Contributions: H.D., S.N.W. and B.M. conceived and designed the experiments; H.D., L.B., M.L., M.G.-S. and O.B. performed the experiments; H.D. and M.G.-S. analyzed the data; H.D., S.N.W. and B.M. wrote the paper. All authors critically reviewed the manuscript.

Funding: The Swiss Nanoscience Institute and the Swiss National Science Foundation (SNSF) have partially financed the research activities within the nanocure project (Project No. 144617). The financial contribution of the Swiss National Science Foundation in the frame of the R'equip initiative (No. 133802) to acquire the micro computed tomography system is also acknowledged.

Acknowledgments: The technical support of F. Schmidli (University of Basel) for the tooth preparation is gratefully acknowledged. Extracted teeth were kindly provided by M. Jakobs (Volkszahnklinik, Basel, Switzerland).

J. Imaging **2018**, *4*, 81

Scanning SAXS experiments were performed at the cSAXS beamline at the Swiss Light Source (proposal numbers 20131216 and 20131056), Paul Scherrer Institut, Villigen, Switzerland.

Conflicts of Interest: The authors declare no conflict of interest. The founding sponsors had no role in the design of the study; in the collection, analyses, or interpretation of data; in the writing of the manuscript, and in the decision to publish the results.

References

1. Marcenes, W.; Kassebaum, N.J.; Bernabé, E.; Flaxman, A.; Naghavi, M.; Lopez, A.; Murray, C.J.L. Global burden of oral conditions in 1990–2010: A systematic analysis. *J. Dent. Res.* **2013**, *92*, 592–597. [CrossRef] [PubMed]
2. Magitot, E. *Treatise on Dental Caries. Experimental and Therapeutic Investigations*; Houghton, Osgood and Company: Boston, MA, USA, 1878.
3. Miller, W.D. Agency of micro-organisms in decay of human teeth. *Dent. Cosmos* **1883**, *25*, 1–12.
4. Braga, M.M.; Mendes, F.M.; Martignon, S.; Ricketts, D.N.J.; Ekstrand, K.R. In vitro Comparison of Nyvad's System and ICDAS-II with Lesion Activity Assessment for Evaluation of Severity and Activity of Occlusal Caries Lesions in Primary Teeth. *Caries Res.* **2009**, *43*, 405–412. [CrossRef] [PubMed]
5. Neuhaus, K.W.; Rodrigues, J.A.; Hug, I.; Stich, H.; Lussi, A. Performance of laser fluorescence devices, visual and radiographic examination for the detection of occlusal caries in primary molars. *Clin. Oral Investig.* **2011**, *15*, 635–641. [CrossRef] [PubMed]
6. Selwitz, R.H.; Ismail, A.I.; Pitts, N.B. Dental caries. *Lancet* **2007**, *369*, 51–59. [CrossRef]
7. Bozdemir, E.; Aktan, A.M.; Ozsevik, A.; Kararslan, E.S.; Ciftci, M.E.; Cebe, M.A. Comparison of different caries detectors for approximal caries detection. *J. Dent. Sci.* **2016**, *11*, 293–298. [CrossRef]
8. Boca, C.; Truyen, B.; Henin, L.; Schulte, A.G.; Stachniss, V.; De Clerck, N.; Cornelis, J.; Bottenberg, P. Comparison of micro-CT imaging and histology for approximal caries detection. *Sci. Rep.* **2017**, *7*, 9. [CrossRef] [PubMed]
9. Abogazalah, N.; Ando, M. Alternative methods to visual and radiographic examinations for approximal caries detection. *J. Oral Sci.* **2017**, *59*, 315–322. [CrossRef] [PubMed]
10. Cheng, R.X.; Shao, J.J.; Gao, X.X.; Tao, C.; Ge, J.Y.; Liu, X.J. Noninvasive Assessment of Early Dental Lesion Using a Dual-Contrast Photoacoustic Tomography. *Sci. Rep.* **2016**, *6*, 9. [CrossRef] [PubMed]
11. Fratzl, P.; Jakob, J.F.; Rinnerthaler, S.; Roschger, P.; Klaushofer, K. Position-resolved small-angle X-ray scattering of complex biological materials. *J. Appl. Crystallogr.* **1997**, *30*, 765–769. [CrossRef]
12. Gupta, H.S.; Roschger, P.; Zizak, I.; Fratzl-Zelman, N.; Nader, A.; Klaushofer, K.; Fratzl, P. Mineralized microstructure of calcified avian tendons: A scanning small angle X-ray scattering study. *Calcif. Tissue Int.* **2003**, *72*, 567–576. [PubMed]
13. Paris, O. From diffraction to imaging: New avenues in studying hierarchical biological tissues with X-ray microbeams (Review). *Biointerphases* **2008**, *3*, FB16–FB26. [CrossRef] [PubMed]
14. Zizak, I.; Roschger, P.; Paris, O.; Misof, B.M.; Berzlanovich, A.; Brenstorff, S.; Amenitsch, H.; Klaushofer, K.; Fratzl, P. Characteristics of mineral particles in the human bone/cartilage interface. *J. Struct. Biol.* **2003**, *141*, 208–217. [CrossRef]
15. Gutierrez, P.; Piña, C.; Lara, V.H.; Bosch, P. Characterization of enamel with variable caries risk. *Arch. Oral. Biol.* **2005**, *50*, 843–848. [CrossRef] [PubMed]
16. Al-Jawad, M.; Streuwer, A.; Kilconey, S.H.; Shore, R.C.; Cywinski, R.; Wood, D.J. 2D mapping of texture and lattice parameters of dental enamel. *Biomaterials* **2007**, *28*, 2908–2914. [CrossRef] [PubMed]
17. Deyhle, H.; White, S.N.; Bunk, O.; Beckmann, F.; Müller, B. Nanostructure of the carious tooth enamel lesion. *Acta Biomater.* **2014**, *10*, 355–364. [CrossRef] [PubMed]
18. Siddiqui, S.; Anderson, P.; Al-Jawad, M. Recovery of Crystallographic Texture in Remineralized Dental Enamel. *PLoS ONE* **2014**, *9*, e108879. [CrossRef] [PubMed]
19. Simmons, L.M.; Al Jawad, M.; Kilcoyne, S.H.; Wood, D.J. Distribution of enamel crystallite orientation through an entire tooth crown studied using synchrotron X-ray diffraction. *Eur. J. Oral Sci.* **2011**, *119*, 19–24. [CrossRef] [PubMed]

20. Tanaka, T.; Yagi, N.; Ohta, T.; Matsuo, Y.; Terada, H.; Kamasaka, K.; To-o, K.; Kometani, T.; Kuriki, T. Evaluation of the distribution and orientation of remineralized enamel crystallites in subsurface lesions by X-ray diffraction. *Car. Res.* **2010**, *44*, 253–259. [CrossRef] [PubMed]

21. Yagi, N.; Ohta, T.; Matsuo, T.; Tanaka, T.; Terada, Y.; Kamasaka, H.; To-o, K.; Kometani, T.; Kuriki, T. Evaluation of enamel crystallites in subsurface lesion by microbeam X-ray diffraction. *J. Synchrotron Radiat.* **2009**, *16*, 398–404. [CrossRef] [PubMed]

22. Shahmoradi, M.; Swain, M.V. Quantitative characterization and micro-CT mineral mapping of natural fissural enamel lesions. *J. Dent.* **2016**, *46*, 23–29. [CrossRef] [PubMed]

23. Rovaris, K.; Matos Ferreira, L.; Oliveira Sousa, T.; Vieira Peroni, L.; Queiroz Freitas, D.; Wenzel, A.; Haiter-Neto, F. Feasibility of micro-computed tomography to detect and classify proximal caries lesions in vitro. *Dent. Res. J.* **2018**, *15*, 123–129.

24. Yagi, N.; Ohta, T.; Matsuo, T.; Tanaka, T.; Terada, Y.; Kamasaka, H.; Kometani, T. A microbeam small-angle X-ray scattering study on enamel crystallites in subsurface lesion. *J. Phys. Conf. Ser.* **2010**, *247*, 012024. [CrossRef]

25. Chien, Y.C.; Burwell, A.K.; Saeki, K.; Fernandez-Martinez, A.; Pugach, M.K.; Nonomura, G.; Habelitz, S.; Ho, S.P.; Rapozo-Hilo, M.; Featherstone, J.D.; et al. Distinct decalcification process of dentin by different cariogenic organic acids: Kinetics, ultrastructure and mechanical properties. *Arch. Oral Biol.* **2016**, *63*, 93–105. [CrossRef] [PubMed]

26. Giannini, C.; Siliqi, D.; Ladisa, M.; Altamura, D.; Diaz, A.; Beraudi, A.; Sibillano, T.; De Caro, L.; Stea, S.; Baruffaldi, F.; et al. Scanning SAXS-WAXS microscopy on osteoarthritis-affected bone—An age-related study. *J. Appl. Crystallogr.* **2014**, *47*, 110–117. [CrossRef]

27. Lewis, R.A.; Rogers, K.D.; Hall, C.J.; Towns-Andrews, E.; Slawson, S.; Evans, A.; Pinder, S.E.; Ellis, I.O.; Boggis, C.R.M.; Hufton, A.P.; et al. Breast cancer diagnosis using scattered X-rays. *J. Synchrotron Radiat.* **2000**, *7*, 348–352. [CrossRef] [PubMed]

28. Falzon, G.; Pearson, S.; Murison, R.; Hall, C.; Siu, K.; Round, A.; Schultke, E.; Kaye, A.H.; Lewis, R. Myelin structure is a key difference in the X-ray scattering signature between meningioma, schwannoma and glioblastoma multiforme. *Phys. Med. Biol.* **2007**, *52*, 6543–6553. [CrossRef] [PubMed]

29. Botta, L.-M.; White, S.N.; Deyhle, H.; Dziadowiec, I.; Schulz, G.; Thalmann, P.; Müller, B. Comparing natural and artificial carious lesions in human crowns by means of conventional hard X-ray micro-tomography and two-dimensional X-ray scattering with synchrotron radiation. *Proc. SPIE* **2016**, *9967*, 99670S.

30. Kind, L.; Stevanovic, S.; Wuttig, S.; Wimberger, S.; Hofer, J.; Müller, B.; Pieles, U. Biomimetic remineralization of carious lesions by self-assembling peptide. *J. Dent. Res.* **2017**, *96*, 790–797. [CrossRef] [PubMed]

31. Deyhle, H.; Schulz, G.; Khimchenko, A.; Bikis, C.N.; Hieber, S.E.; Jaquiery, C.; Kunz, C.; Müller-Gerbl, M.; Hoechel, S.; Saxer, T.; et al. Imaging tissues for biomedical research using the high-resolution micro-tomography system nanotom®m. *Proc. SPIE* **2016**, *9967*, 99670Q.

32. Dziadowiec, I.; Beckmann, F.; Schulz, G.; Deyhle, H.; Müller, B. Characterization of a human tooth with carious lesions using conventional and synchrotron radiation-based micro computed tomography. *Proc. SPIE* **2014**, *9212*, 92120W.

33. Bunk, O.; Bech, M.; Jensen, T.H.; Feidenhans'l, R.; Binderup, T.; Menzel, A.; Pfeiffer, F. Multimodal X-ray scatter imaging. *New J. Phys.* **2009**, *11*, 123016. [CrossRef]

34. Lloyd, S.P. Least square quantization in PCM. *IEEE Trans. Inf. Theory* **1982**, *28*, 129–137. [CrossRef]

35. Müller, B.; Beckmann, F.; Huser, M.; Maspero, F.; Szekely, G.; Ruffieux, K.; Thurner, P.; Wintermantel, E. Non-destructive three-dimensional evaluation of a polymer sponge by micro-tomography using synchrotron radiation. *Biomol. Eng.* **2002**, *19*, 73–78. [CrossRef]

36. Deyhle, H.; Bunk, O.; Müller, B. Nanostructure of healthy and caries-affected human teeth. *Nanomed. Nanotechnol. Biol. Med.* **2011**, *7*, 694–701. [CrossRef] [PubMed]

37. Sui, T.; Sandholzer, M.A.; Baimpas, N.; Dolbnya, I.P.; Landini, G.; Korsunsky, A.M. Hierarchical modelling of elastic behaviour of human enamel based on synchrotron diffraction characterisation. *J. Struct. Biol.* **2013**, *184*, 136–146. [CrossRef] [PubMed]

38. Schulz, G.; Waschkies, C.; Pfeiffer, F.; Zanette, I.; Weitkamp, T.; David, C.; Müller, B. Multimodal imaging of human cerebellum—Merging X-ray phase microtomography, magnetic resonance microscopy and histology. *Sci. Rep.* **2012**, *2*, 826. [CrossRef] [PubMed]

39. Stalder, A.; Ilgenstein, B.; Chicerova, N.; Deyhle, H.; Beckmann, F.; Müller, B.; Hieber, S.E. Combined use of micro computed tomography and histology to evaluate the regenerative capacity of bone grafting materials. *Int. J. Mater. Res.* **2014**, *105*, 679–691. [CrossRef]

40. Deyhle, H.; Dziadowiec, I.; Kind, L.; Thalmann, P.; Schulz, G.; Müller, B. Mineralization of early stage carious lesions in vitro—A quantitative approach. *Dent. J.* **2015**, *3*, 111–122. [CrossRef] [PubMed]

41. Ten Cate, J.M.; Dundon, K.A.; Vernon, P.G.; Damato, F.A.; Huntington, E.; Exterkate, R.A.M.; Wefel, J.S.; Jordan, T.; Stephen, K.W.; Roberts, A.J. Preparation and measurement of artificial enamel lesions, a four-laboratory ring test. *Car. Res.* **1996**, *30*, 400–407. [CrossRef] [PubMed]

42. Jensen, T. *Refraction and Scattering Based X-ray Imaging*; University of Copenhagen: Copenhagen, Denmark, 2010.

43. Lautensack, J.; Rack, A.; Redenbach, C.; Zabler, S.; Fischer, H.; Graber, H.G. In situ demineralisation of human enamel studied by synchrotron-based X-ray microtomography—A descriptive pilot-study. *Micron* **2013**, *44*, 404–409. [CrossRef] [PubMed]

44. Arends, J.; Christoffersen, J. The nature of early caries lesions in enamel. *J. Dent. Res.* **1986**, *65*, 2–11. [CrossRef] [PubMed]

45. Pfeiffer, F.; Bech, M.; Bunk, O.; Kraft, P.; Eikenberry, E.F.; Brönnimann, C.; Grünzweig, C.; David, C. Hard-X-ray dark-field imaging using a grating interferometer. *Nat. Mater.* **2008**, *7*, 134–137. [CrossRef] [PubMed]

46. Revol, V.; Jerjen, I.; Kottler, C.; Schutz, P.; Kaufmann, R.; Luthi, T.; Sennhauser, U.; Straumann, U.; Urban, C. Sub-pixel porosity revealed by X-ray scatter dark field imaging. *J. Appl. Phys.* **2011**, *110*, 5. [CrossRef]

47. Jensen, T.; Bech, M.; Zanette, I.; Weitkamp, T.; David, C.; Deyhle, H.; Feidenhans'l, R.; Pfeiffer, F. Directional X-ray dark-field imaging of strongly ordered systems. *Phys. Rev. B* **2010**, *82*, 214103. [CrossRef]

Journal of
Imaging

MDPI

Article

Non-Destructive Testing of Archaeological Findings by Grating-Based X-Ray Phase-Contrast and Dark-Field Imaging

Veronika Ludwig [1,*], Maria Seifert [1], Tracy Niepold [2], Georg Pelzer [1], Jens Rieger [1], Julia Ziegler [3], Thilo Michel [1] and Gisela Anton [1]

[1] Erlangen Centre for Astroparticle Physics, Friedrich-Alexander-University Erlangen-Nuremberg, Erwin-Rommel-Strasse 1, 91058 Erlangen, Germany; maria.seifert@fau.de (M.S.); georg.pelzer@fau.de (G.P.); jens.rieger@fau.de (J.R.); thilo.michel@fau.de (T.M.); gisela.anton@fau.de (G.A.)
[2] Bayerisches Landesamt für Denkmalpflege, Dienststelle Bamberg, Schloss Seehof 1, 96117 Memmelsdorf, Germany; Tracy.Niepold@blfd.bayern.de
[3] Deutsches Ledermuseum Offenbach, Frankfurter Straße 86, 63067 Offenbach, Germany; Ziegler@ledermuseum.de
* Correspondence: veronika.ludwig@fau.de; Tel.: +49-9131-85-27124

Received: 28 February 2018; Accepted: 10 April 2018; Published: 14 April 2018

Abstract: The analysis of archaeological findings reveals the remaining secrets of human history. However, it is a challenging task to investigate and simultaneously preserve the unique remains. Available non-destructive examination methods are limited and often insufficient. Thus, we considered X-ray grating interferometry as a non-destructive and advanced X-ray imaging method to retrieve more information about archaeological findings. In addition to the conventional attenuation image, the differential phase and the dark-field image are obtained. We studied the potential of the scattering-sensitive dark-field and the phase-shift sensitive differential phase image to analyse archaeological findings. Hereby, the focus lies on organic remnants. Usually, the organic materials have vanished due to decomposition processes, but the structures are often preserved by mineralisation and penetration of corrosion products. We proved that the combination of the attenuation and the dark-field image in particular, enables a separation of structural properties for fabric remnants. Furthermore, we achieved promising results for the reconstruction of sub-pixel sized fibre orientations of woven fabric remnants by employing the directional dark-field imaging method. We conclude from our results that a further application of X-ray dark-field imaging on wet organic findings and on the distinction of different types of organic remnants at archaeological findings is promising.

Keywords: archaeological findings; non-destructive materials testing; X-ray phase-contrast imaging; Talbot-Lau interferometry; directional dark-field imaging; sub-pixel fibre orientation; organic remnants; fabric remnants

---

1. Introduction

Archaeological findings are the remaining traces of human history. By investigation of these ancient objects we get an insight in the daily life and human habits of bygone eras. As long as the archaeological findings are embedded in their original environment, they constitute unique evidence of the past. The discovery of archaeological sites, e.g., due to building projects, initiates a process of recovery, conservation and restoration. The primary objective is the preservation of the ancient remnants of cultural heritage. Cultural heritage management plays an important role in managing the care of archaeological sites, supporting excavations and the following conservation process of archaeological findings. In addition to the importance as man-made artefacts, archaeological findings may reveal answers concerning dating, provenance, the way of deposition, the context of other findings and the former surrounding like ritual wrapping with textiles or contact to clothes of the deceased. After excavation, archaeological findings remain as unique witnesses of the archaeological site. Thus, one of the tasks of archaeological and conservation science is to investigate these remains very carefully and non-destructively [1,2]. Therefore, X-ray imaging is a favoured analysis method. The application of X-rays on archaeological findings already happens by digital radiography, computed tomography [3,4], X-ray fluorescence or X-ray diffraction methods [5]. Investigating metal artefacts by X-ray imaging is a common way to get information about the material, the dimension and the preservation status of the enclosed finding. This method plays an important role if artefacts are lifted with the surrounding soil (block excavation) due to a fragile preservation status or in case of assumed organic remains. We propose the application of X-ray differential phase-contrast and dark-field imaging on archaeological findings as an extended X-ray imaging method.

The grating-based Talbot-Lau interferometer setup for phase-contrast imaging was introduced in 2006 by Pfeiffer et al. [6]. An enhancement of the conventional X-ray imaging is given by the access to the real part of the complex refractive index of an object material. The method enables us to get more information about the inner structure and material properties by providing two additional images: the differential phase image and the dark-field image. The differential phase image reveals gradients in the real part of the refractive index, which shows better contrast for imaging light elements [7]. In addition, the X-ray dark-field image can be used for the inspection of porosity [8] and visualisation of structures below the spatial resolution of the detector [9]. The dark-field signal is mainly based on small angle X-ray scattering (SAXS) caused by variations in the electron density of an inhomogeneous sample [10,11]. As a further analysis tool, directional X-ray dark-field imaging is used in order to reconstruct fibre orientations of micrometer-sized aligned structures which are not resolvable directly [12–14]. The method is applicable for the analysis of bones [15–17] or the characterisation of fibre alignments in polymers [18–20]. An extension of directional dark-field imaging to a 3D analysis of fibre orientations is done by tensor tomography [21–23].

Applying this novel interferometric X-ray imaging method on archaeological findings could improve the non-destructive analysis compared to conventional X-ray imaging. In this paper the investigation of archaeological findings by grating-based X-ray phase-contrast and dark-field imaging with a high-energy setup is presented. In particular, X-ray dark-field imaging proved to be a promising examination method for archaeological findings. First, we analyse X-ray differential phase-contrast and dark-field imaging regarding the differentiation of structural properties. In particular, the analysis of mineralised organic remains will be addressed. Second, we examine the influence of soil particles adhered to archaeological findings. In this regard, we additionally evaluate the applicability of X-ray phase-contrast imaging as an extended X-ray prospection method to obtain more information on enclosed findings in block excavations, especially with the focus on detecting mineralised organic remnants. Third, we employ the directional dependence of the X-ray dark-field to reveal fibrous structures and to reconstruct structure orientations in samples containing mineralised fabric remnants. Finally, we discuss the advantages and occurring problems of applying X-ray grating interferometry on archaeological findings.

2. Materials and Methods

2.1. Grating-Based X-ray Phase-Contrast Imaging

X-ray phase-contrast imaging provides access to both the imaginary and the real part of the complex refractive index

$$n = 1 - \delta + i\beta .$$ (1)

In common X-ray imaging, the attenuation coefficient $\mu = 4\pi\beta/\lambda$, with λ being the wavelength of the interacting photons, comprises the complete image information. By use of a so-called Talbot-Lau interferometer also the phase shift $\Phi = 2\pi\delta/\lambda$ is accessible. The setup of a Talbot-Lau interferometer is sketched in Figure 1. In this work we used a setup which consists of a conventional X-ray tube of low brilliance, two absorption gratings, one phase grating and a flat-panel X-ray detector. The basic concept relies on the Talbot effect. The phase grating G1 imprints a phase-shift onto the incoming wavefront. In certain so-called fractional Talbot distances behind the grating, an intensity pattern with maximal contrast can be measured, which reproduces the periodic structure of G1 [24,25]. Due to the extended focal spot of the X-ray source, a source grating G0 is needed. The idea is to generate many mutually incoherent slit sources, such that the resulting intensity patterns created by each single slit overlap constructively [6]. As the detector pixel size is larger than the period of the interference pattern, it is not possible to resolve the pattern directly. The third grating G2 with absorbing lamellae has the same period as the Talbot pattern and is placed at the position of its appearance. By laterally shifting G2 perpendicular to the beam path and to the grating lines, the intensity modulation is sampled. Thereby, the G2 grating is moved in fractions of its period p_2. This process is called phase-stepping. At every phase-step, an image is acquired, which leads to a sinusoidal intensity curve in each pixel as a function of the phase-step position [7]. This phase-stepping curve is measured once for a reference and once for an object measurement. The intensity variation I for different phase-step positions x is given by

$$I(x) = I_0 + A \cdot \sin\left(\frac{2\pi}{p_2}x + \varphi\right) . \tag{2}$$

The curve parameters mean intensity I_0, amplitude A and phase φ are reconstructed for each pixel either by Fourier analysis or least-square fit. The contrast or visibility of the curve is defined as $V = A/I_0$. The obtained parameters of the measured reference curve and object curve are used to calculate the three mentioned images attenuation Γ, differential phase $\Delta\Phi$ and dark-field Σ:

$$\Gamma := -\ln\left(\frac{I_{0,\text{obj}}}{I_{0,\text{ref}}}\right) , \qquad \Delta\Phi := \varphi_{\text{obj}} - \varphi_{\text{ref}} , \qquad \Sigma := -\ln\left(\frac{V_{\text{obj}}}{V_{\text{ref}}}\right) . \tag{3}$$

The conventional attenuation image reveals the amount of photons absorbed by passing through the sample. The differential phase image describes the first derivative of the phase shift that the sample adds to the incoming X-ray wave front. Thus, edges are enhanced and differences between light elements get apparent [26]. The dark-field image originates from the reduction of contrast in the Talbot pattern. Hereby, a visualisation of structures below the spatial resolution of the imaging system is achieveable [9,10]. Additionally, we introduce the normalised scatter image, which follows the definition of the R-value [27]: $R = \Sigma/\Gamma$. The normalised scatter image is independent of the thickness and particularly enhances scattering properties. In this regard, plotting the dark-field values pixelwise over the attenuation values in a so-called scatter plot enables a differentiation of materials by both attenuation and scattering properties [28].

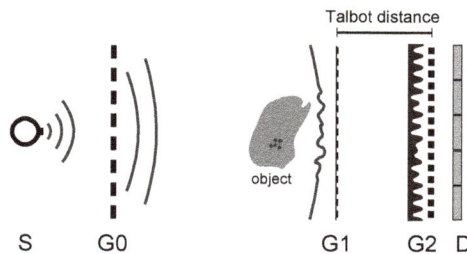

Figure 1. Sketch of a Talbot-Lau interferometer with an X-ray source S, the source grating G0, an object, the phase grating G1, the analyser grating G2 and a flat-panel X-ray detector D. The pattern in front of G2 illustrates the Talbot intensity pattern, which is produced by G1 and distorted by the object.

2.2. Directional X-ray Dark-Field Imaging

Dark-field images include signals which occur due to scattering at micrometer-sized structures, which can not be resolved directly [10,29,30]. A typical Talbot-Lau interferometer uses line gratings, which generates sensitivity to the wave front gradient in the direction perpendicular to the grating bars. Thus, scattering and phase shifts along the direction of the grating lines have no influence on the contrast of the Talbot pattern. The directional dependence of the dark-field image holds the capability of retrieving information about the angular variations of the local scattering power of a sample [12,14]. For that purpose, the object is rotated around the beam axis and at least three images at different angular orientations of the object are acquired. The periodical variation of the dark-field signal Σ depending on the angular orientation α of the sample with respect to the grating bars follows [12]

$$\Sigma = A_\alpha \cdot \cos(2 \cdot (\alpha + \Phi_\alpha)) + I_\alpha \quad . \tag{4}$$

By extracting the parameters A_α, Φ_α and I_α information about the main fibre orientation $\alpha = \Phi_\alpha$, the average scattering strength I_α, the anisotropic part of the scattering strength A_α and the degree of anisotropy $V_\alpha = A_\alpha / I_\alpha$ is obtainable [16,20].

2.3. The Experimental Setup

The measurements of the archaeological findings were performed using a laboratory-based X-ray phase-contrast setup with a MEGALIX CAT Plus 125 medical X-ray tube (Siemens, Munich, Germany). X-ray spectra of maximum peak voltages from 60 to 120 kVp were generated by a tungsten anode and 0.3 mm copper filtration. For all measurements, nine phase-steps over one period were acquired with a Dexela 1512 X-ray flat panel detector (PerkinElmer, Waltham, MA, USA) with 74.8 μm pixel size in 2×2 binning mode. The grating parameters and further setup details are comprised in Table 1. The used grating set is a result of an optimisation process, published in [31]. The gratings are made of gold and are fabricated by the Karlsruhe Nano Micro Facility/Karlsruhe Institute of Technology using deep X-ray lithography [32]. The archaeological findings were fixed to a rotation plate in order to retrieve images at different angular positions by rotating the sample around the beam axis. In addition, the sample tray could be moved in the object plane to cover larger samples and to acquire reference images.

In Figure 2 on the left-hand side the (monochromatic) simulated visibilities, achievable for the used setup parameters (grating details and distances, see Table 1) for energies between 25 to 120 keV, are shown in a so-called visibility curve. Further details about the used wave-field simulation can be found in [33]. The energy spectra for 60 and 120 kVp (provided by [34]) are overlaid to the same plot. Additionally, the shown spectra include a 0.3 mm copper filtration and the detection efficiency of 600 μm CsI. On the right-hand side the measured reference visibilities in each pixel are depicted in a visibility map for the applied spectra of 60, 80, 100 and 120 kVp. Since the visibility measured by using a broad X-ray spectrum is a weighted average of the contributing monoenergetic visibilities, the shape of the spectrum is significant. Therefore, the spectrum must fit well to the monoenergetic visibility curve depicted in Figure 2 in blue in order to achieve a high reference visibility in the polychromatic case. We obtain the best visibilities for energies around 45 keV, leading to a high measured reference visibility of about 37 % for the 60 kVp spectrum. For the further applied spectra of 80, 100 and 120 kVp the reference visibility drops to 22%, 18% and 16%, respectively. For the dark-field image the resulting object visibility compared to the reference visibility is decisive. In case of highly absorbing materials in a sample, beam hardening effects also play an important role for the dark-field signal generation [35,36]. Cutting the lower energies of the spectrum, even higher object visibilities than reference visibilities can be reached in some cases.

Table 1. Parameters of the used gratings (left) and distances (right). In this configuration a magnification factor of 1.87 is obtained.

Grating	Period (μm)	Height (μm)	Duty Cycle	Distances (cm)			
G0	13.31	200	0.5	$d_{G0\text{-}G1}$	90	$d_{focus\text{-}G0}$	16
G1	5.71	6.3	0.3	$d_{G1\text{-}G2}$	68	$d_{G0\text{-}object}$	79.5
G2	10	200	0.5	$d_{G2\text{-}detector}$	6.1		

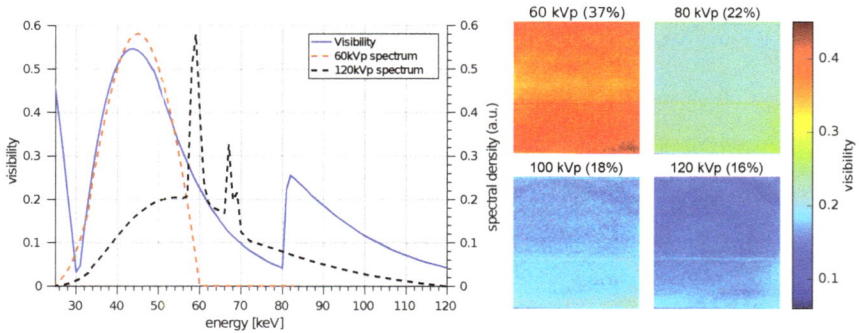

Figure 2. Simulated visibility (**left**) shown in blue for the used setup for energies from 25 to 120 keV, referring to the left vertical axis. The dashed lines represent the applied 60 kVp (red) and 120 kVp (black) spectrum, filtered with 0.3 mm copper and an included detection efficiency of 600 μm CsI, referring to the right axis. The measured reference visibility maps are shown for 60, 80, 100 and 120 kVp spectra (**right**). The retrieved average visibility of the polychromatic measurement is denoted in the brackets.

2.4. Details about the Examined Test and Archaeological Samples

The test samples and the archaeological findings were provided by the Bavarian State Office for Heritage Management (BLfD, "Bayerisches Landesamt für Denkmalpflege"). A photography of the test samples is shown in Figure 3. The test samples consist of different types of fabric, comprising woven linen and hemp, single woolen threads dyed with different natural dyestuffs, a piece of cowhide leather and a replicated, non-ferrous metal fibula featuring a bird motif with some woolen threads tied to it. The organic test samples are typical materials expected to occur in a mineralised form attached to metallic findings.

Moreover, we analysed archaeological findings chosen from two early medieval cemeteries (Regensburg-Burgweinting and Baar-Oberbaar) which date back to the sixth century after Christ. The buried persons were pagans to whom the relatives gave their personal belongings for representing their status in the beyond. Therefore, a variety of metal artefacts — weaponry, cloth fasteners or metal applications — is preserved. Metallic parts of findings are preserved due to the processes of corrosion organic materials. Normally, organic materials would have vanished because of decomposition. However, if organic substances are adherent to metallic materials, they are penetrated by soluble corrosion products. A partial or full transformation of the organic material or a negative imprint of the organic structure in the corrosion layer follows. This enables an analysis of textiles and other organic materials which are important for context-related information, e.g., on cultural history or on trade links.

In this work we investigated an iron fibula, an iron strap-end and a bird fibula which are depicted in Figure 4. The iron fibula and the strap-end have a prominent iron core. The bird fibula is made of non-ferrous metal and is gold-plated. The samples represent typical findings of different metals and mineralised organic remains at different stages of processing. The iron fibula shows a fragment of mineralised linen fabrics at the inner side. The nearer the fabric remnants are to the iron parts, the higher is the proportion of metallic elements in the fabric part. Investigations shall be undertaken to determine whether it is feasible to separate different states of mineralised fabrics by the dark-field and the normalised scatter image. While the fibulas are fully exposed, the iron strap-end differs from the other samples by demonstrating soil remnants on the outer parts. Below the layer of soil, organic remnants are presumed. It is necessary to evaluate how much soil particles affect the analysis of X-ray phase-contrast and dark-field images. The bird fibula also represents an interesting sample regarding mineralised fabric remnants, which are visible around the pin. The question is raised as to whether spinning directions in mineralised fabrics are reconstructable. This may be enabled by applying directional X-ray dark-field imaging.

Figure 3. Photography of the examined test samples. Two different types of woven fabrics made of linen and hemp are shown in the top left corner. On the right side lies a piece of cowhide leather. Between the cloth and the leather, a replicated bird fibula is placed. The bottom row shows strands of dyed wool.

Figure 4. Photography of the examined archaeological findings: Iron fibula, strap-end and bird fibula. Photographies provided by the Bavarian State Office for Heritage Management (BLfD), Tracy Niepold, Julia Ziegler, Helmut Voss.

3. Results

3.1. Pre-Measurements on Test Samples

One aim of the investigation of archaeological findings is the detection and analysis of mineralised fabric remnants. These are potentially visible by dark-field imaging due to their scattering properties compared with non-fibrous materials. In order to test the potential of dark-field and differential phase images for visualisation of organic materials, different test samples (photography Figure 3) are imaged. In Figure 5, the results of the attenuation, differential phase, dark-field and normalised scatter image are shown. In the bottom row, a detailed view of the woven fabrics and some threads is given for the chosen region of interest which is marked by a red rectangle. The applied energy spectrum for the reference materials was 60 kVp and the images were acquired with 7.5 mAs per phase-step. A lower energy spectrum would be more suitable for these reference materials. However, the choice for the used (higher) energy spectrum is motivated by the aspect of imaging iron-based archaeological findings with attached fabric remnants. Because of the low attenuation of woven materials at these energies, the contrast in the attenuation image is quite weak. The differential phase image shows the texture of the woven fabrics fairly well. This is better visible in the detailed view, which reveals the threads. However, the contrast is low. The periodic structure over the whole differential phase image occurs due to Moiré artefacts which can occur due to slight mechanical instabilities during the phase-stepping procedure. These get more obvious due to the small range of depicted differential phase values from -0.06 to 0.06. In the following analysis we will particularly focus on the dark-field image and the normalised scatter image, which offer the best contrast for organic materials, even at the chosen 60 kVp energy spectrum. All types of fabrics of the test samples are clearly visible, including structural information. A defect in the piece of leather on the right side is revealed. Some lines presenting cracks on the surface of the leather are also apparent. The dark-field shows each sling of the threads even those lying on the leather or the woven fabrics. Hence, in particular, the dark-field image and the normalised scatter image have the potential for investigating organic remains in archaeological findings.

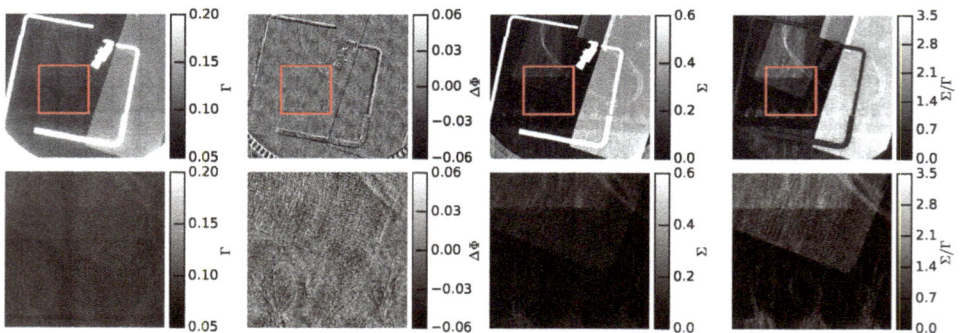

Figure 5. From left to right: Attenuation, differential phase, dark-field image and normalised scatter image of the test samples from Figure 3, taken at 60 kVp and 7.5 mAs per phase-step. The region marked by a red square is shown in detail in the bottom row.

3.2. X-ray Phase-Contrast Imaging of Archaeological Findings

3.2.1. Normalised Scatter Image and Scatter Plot of an Iron Fibula

In Figure 6 attenuation, differential-phase and dark-field images of the iron fibula are shown. The images were acquired at 80 kVp and 22.5 mAs per phase-step. The highly absorbing part, clearly visible in the attenuation image, represents the iron-based bow of the fibula. Due to beam hardening the mean energy of the spectrum is strongly shifted to higher energies resulting in this case in a lower visibility. In addition, the highly absorbing object regions lead to low photon statistics and a noise level surpassing the visibility of the phase-stepping curve. This leads to saturated dark-field values and no reasonable reconstruction of phase information in the iron part of the fibula is possible. Between the pin rest and iron bow of the fibula, some mineralised fabric remains are preserved. These are better visualised in the dark-field and the differential phase image. The structure of the mineralised textile is clearly enhanced by dark-field imaging. Thus, interferometric X-ray imaging can be utilized for further analysis of archaeological findings, especially for non-metal parts of a sample. The differential phase image enhances edges.

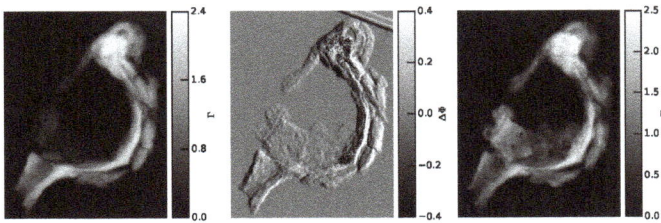

Figure 6. Attenuation (**left**), differential phase (**middle**) and dark-field image (**right**) of the iron fibula from Figure 4, taken at 80 kVp and 22.5 mAs per phase-step.

To enhance the scattering parts in the iron fibula, the thickness-independent normalised scatter image is viewed (Figure 7, left). The fabric remnants at the inner side of the fibula and also at the outer bottom side are highlighted. The iron part shows sharper contours and a uniform surface in separated segments. Next to the normalised scatter image, the scatter plot for the iron fibula is depicted in Figure 7. The coloured regions in the plot were chosen manually by searching frequent occurrences of certain Σ-Γ-values. The chosen colours in the plot refer to different regions, which are shown in the image on the right. Blue, green and yellow illustrate the mineralised fabrics, which are clearly enhanced by the normalised scatter image. In red and orange the iron part of the fibula and the transitions from iron to mineralised fabric are visible. The fabric remnants nearest to the strong iron core (orange and yellow) reach higher attenuation values than the blue and green region. In particular, the high dark-field values and the low attenuation values are characteristic for the blue area. The degree of mineralisation might be at its lowest, or the kind of fabric differs from the surrounding fabric remains. The appropriate dark-field and attenuation pairs lie on a straight line (blue) on the left-hand side in the scatter plot, indicating similar structural and material properties of the blue region. This shows that analysing the normalised scattering image and the scatter plot might be a helpful tool to examine archaeological findings, in particular, regarding organic remains.

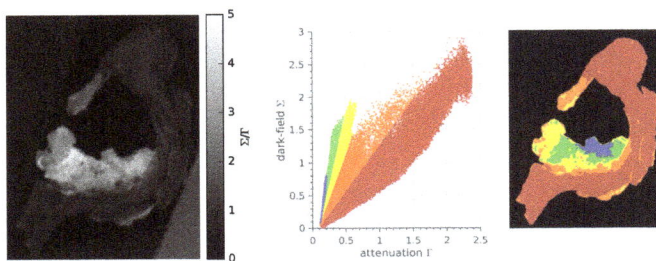

Figure 7. Normalised scatter image of the iron fibula, acquired at 80 kVp and 22.5 mAs per phase-step (**left**). The scatter plot and image (**right**) show the abundance of measured Γ- and Σ-values for the iron fibula with the colours referring to different regions in the Γ-Σ-plain, respectively in the fibula image on the right.

3.2.2. Influence of Soil on Dark-Field Images of a Strap-End

Soil leftovers are usually present, even if the archaeological finding was not taken out by a block excavation. Therefore, we examined how differential phase-contrast and dark-field imaging deals with scattering signals caused by soil particles. This is an important aspect for the applicability regarding the prospection of block excavations.

As a preliminary investigation, a sponge, which causes a clear dark-field signal due to the air-sponge boundaries, is imaged while being placed on varying heights of dry soil. The averaged resulting attenuation and dark-field signals with and without a sponge placed on an increasing amount of soil particles are plotted in Figure 8 on the left-hand side. Note that the thickness of soil shown on the *x*-axis in the plot has been roughly estimated by the resulting attenuation images. For that purpose the absorption coefficient of concrete obtained from NIST [37] has been assumed. The approximately measured thicknesses for each soil layer was about twice the value in the plot, but was taken to be imprecise because of the unknown degree of compression and variable soil components. Thus, the shown soil layer heights in the plot do not represent quantitative values, but qualitatively demonstrate the influence of the increasing amount of soil particles. The sponge itself (without soil) causes a clear dark-field signal. By adding soil, the contrast between sponge structure and soil particles decreases rapidly. Even a small layer of soil already results in high enough dark-field signals to conceal the signal caused by the sponge. From a soil thickness of around 5 mm and over, a differentiation of soil and soil with sponge is hardly possible. Moreover, the dark-field signal saturates quite fast with increasing soil layer height. This makes it impossible to gain information about objects which are covered by soil. The plot on the right in Figure 8 shows the effect of adding water to the soil. Due to additional attenuation by the water, the attenuation signal increases. However, the dark-field signal clearly decreases. The scattering properties are reduced, because of less scattering at water-soil-boundaries compared to scattering at air-soil-boundaries. Thus, the influence of disturbing soil particles could be reduced slightly by moisturising covering soil layers. However, for the investigation of archaeological findings, moisturising the surrounding soil is no applicable technique. The moisture may increase corroding processes and may cause harmful swelling and shrinkage processes to the adhered organic materials. We conclude that X-ray phase-contrast imaging is not feasible for an analysis of block excavations. However, small remaining layers of soil on an almost fully exposed sample could possibly be handled.

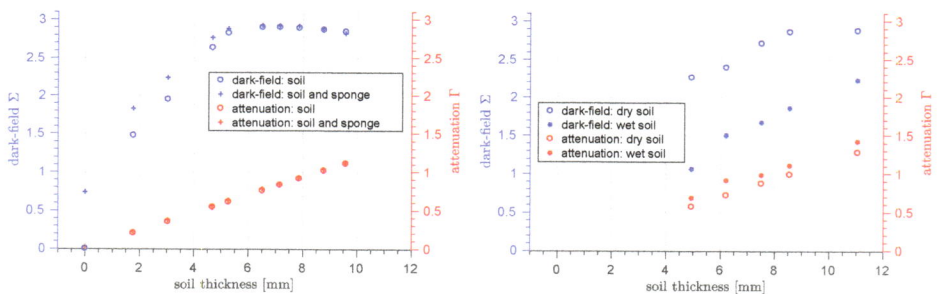

Figure 8. Attenuation (red) and dark-field values (blue) obtained by different heights of soil with two sponges (plus) and without (circle) are plotted over the thickness of soil (**left**). The first plus signs (soil thickness equal to zero) represent the dark-field and attenuation signal that the sponges generate. The effect on the attenuation (red) and dark-field values (blue) of dry soil (circle) and wet soil (filled circle) is compared for increasing heights of soil (**right**).

Thus, the strap-end, shown in the photograph in Figure 4, is examined. It is mainly composed of iron and, in addition, leftovers of soil cover the bottom and the lower part of the archaeological finding. Below the layer of soil, organic remains are presumed. As shown before, soil particles induce scattering and hence, a strong dark-field signal, which rapidly saturates and conceals further structures. The resulting attenuation, differential phase, dark-field and the normalised scattering image of the strap-end are presented in Figure 9. Here, 60 mAs were used at a 60 kVp spectrum with the highest reference visibility of 37 %. The use of a higher energy spectrum (up to 120 kVp) has been tested. The results are comparable; only slight differences are recognisable with the images at 60 kVp providing little more detail. As a result of the existing soil particles in combination with the occurrence of iron, high dark-field values are caused, especially in the lower part of the strap-end, probably hiding the subjacent structures. The attenuation image indicates a relatively homogeneous material composition with a high percentage of metal. In contrast to the attenuation image, the middle part differs from the upper and lower part of the strap-end in the dark-field image. Assuming a similar influence of the metal over the whole strap-end, the remaining soil particles are responsible for the high dark-field values in the lower part of the strap-end. The higher amount of scattering particles, covering many parts of the strap-end, is also apparent by comparing the differential phase image of the here shown strap-end with the result for the iron fibula of Figure 6. More regions of the strap-end look noisy in the differential phase image, which is caused by scattering to a larger degree than by beam hardening. This is especially observable on the bottom left and on the outer bottom part of the strap-end. Here, the attenuation is comparably small, but the differential phase shows a noisy behaviour, nevertheless. Due to the soil particles, the visibility is reduced to such a large degree that the phase retrieval is not possible. In the normalised scatter image, a circular structure emerges more clearly. We assume it could be a rivet, which is common at strap-ends. At the edges and in particular in the lower outer region on the left, mainly soil remnants are present, which lead to a strong normalised scatter signal. In the lower region of the strap-end, which is covered by soil, some slight variations are recognisable in the normalised scatter image. The further analysis after removing the soil layer from the lower part of the strap-end revealed a piece of leather without further structuring. Whether the signal variations occur due to this organic remnant could not be verified.

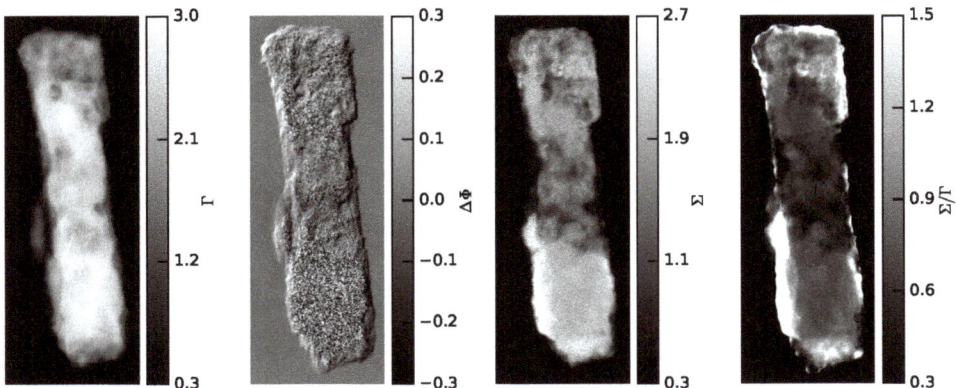

Figure 9. From left to right: Attenuation, differential phase, dark-field and normalised scatter image of the strap-end from Figure 3, acquired at 60 kVp and 60 mAs per phase-step.

3.2.3. Directional X-ray Dark-Field Imaging of a Bird Fibula

In a process of corrosion, the original organic-based textiles vanish by decomposition, but their structure is mainly preserved in the resulting mineralised remnants of the fabrics. This enables an orientation-dependent analysis by directional dark-field imaging. In Figure 10, the obtained attenuation, differential phase, dark-field and normalised scatter image of the bird fibula from Figure 4 are presented. The images were acquired at 120 kVp and 2 mAs per phase-step. On the backside of the fibula, some mineralised textile remains are preserved next to the pin rest. These are visible, particularly in the dark-field image and even more enhanced in the normalised scatter image. The contrast in the attenuation image of this region is rather low. In the differential phase image, a slightly visible structure is recognisable. Images were acquired at four different orientations of the fibula by rotation around the beam axis. The resulting images are used to apply the directional dark-field imaging method by a pixelwise analysis. The resulting degree of anisotropy and fibre orientation is depicted in Figure 11. The preferred fibre orientations Φ_α are presented as lines, pointing in the direction of the main structure alignment and printed on the colour-coded degree of anisotropy, which indicates the amount of aligned fibres with orientation Φ_α. The part containing fabric remnants demonstrates a clear directional dependence caused by the mineralised woven material around the pin on the backside of the fibula (compare detailed view in Figure 4). The eye and the tail of the bird also seem to show a directional dependence. In this case, the variation of the dark-field signal during rotation in one analysed pixel probably occurs due to notches and edges. In the process of matching the images of each angular position, it is not possible to obtain a perfect pixelwise overlap. Therefore, in combination with the small number of angular positions, edges create the impression of a similar variation in the dark-field signal strength during rotation as that which would be caused be true fibre alignments. Those variations, however, are casual and could be excluded by measuring a higher number of angular positions of the sample. The analysis of directional dark-field imaging proved here to be a successful method for investigating mineralised fibre structures. An alignment of fibrous structures, which are not visible in any of the three images, could be revealed.

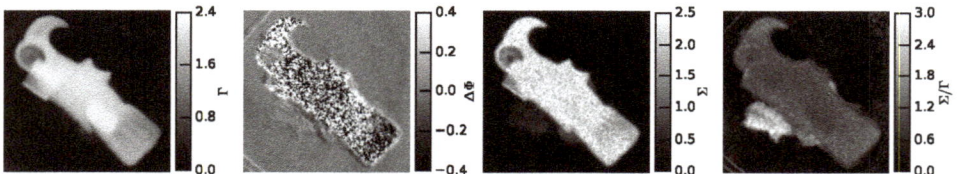

Figure 10. From left to right: Bird fibula shown in attenuation, differential phase, dark-field and normalised scatter image, acquired at 120 kVp and 2 mAs per phase-step.

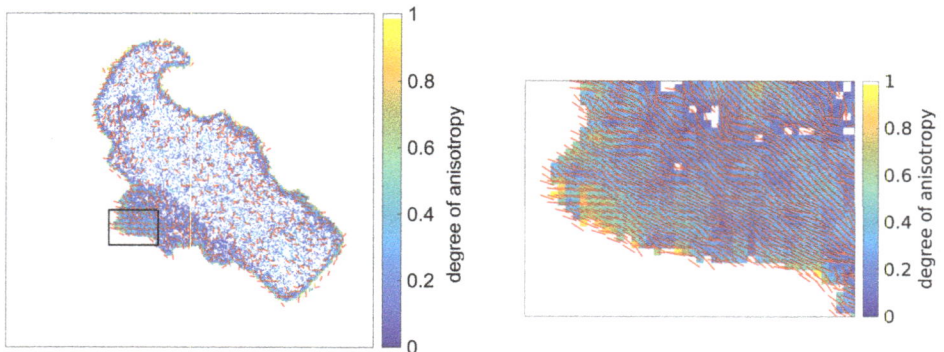

Figure 11. Reconstruction of fibre orientations (**left**) caused by the mineralised textile structures at the pin of the bird fibula (photography in Figure 4). A zoom on the region in the left figure marked with a black rectangle is shown (**right**). The red lines point in the main direction of aligned structures and are plotted on the colour-coded degree of anisotropy.

4. Discussion and Conclusions

We employed X-ray differential phase-contrast and dark-field imaging, providing an extended X-ray imaging method, on the non-destructive investigation of archaeological findings. The high amount of metallic parts and soil remains proved to be the main problems in the analysis of the examined samples. However, we obtained additional information for fully exposed findings by the dark-field and normalised scatter image. Moreover, we achieved promising results for the analysis of sub-pixel sized fibre orientations of mineralised fabric remnants by directional X-ray dark-field imaging. Our pre-measurements on test samples of different organic materials demonstrated such that the dark-field image and the normalised scatter image are suitable to visualise fabrics and leather better than the conventional attenuation image. The investigated archaeological findings comprised an iron fibula with mineralised fabric remnants, a strap-end partly covered with soil remains and a bird fibula with well preserved mineralised woven textile structures.

One problem we faced in the analysis of the archaeological findings is the high percentage of metallic parts. High absorbing materials cause beam hardening, which we have obtained in the results of the differential phase and dark-field image of the iron fibula and the strap-end. Higher spectrum energies could be helpful to reduce the influence, but for that purpose, optimised setups have to be developed. The main challenge here is given by obtaining high reference visibilities for high photon energies, which is mainly limited by the achievable height of the analyser grating.

Nevertheless, we demonstrated that an investigation of fully exposed archaeological findings with X-ray dark-field imaging provides additional information, in particular regarding fabric remnants. We showed that it is feasible to separate different materials due to thickness-independent scattering properties. For that purpose, we employed the normalised scatter image and the scatter plot in order to differentiate regions of different degrees of mineralisation and iron parts of the iron fibula. A mapping of regions differing in the degree of mineralisation proved feasible.

Furthermore, we showed that even small layers of soil disturb the obtainable dark-field signals significantly. A differentiation between scattering structures and the occurrence of scattering due to soil particles is hardly possible. This means that X-ray dark-field imaging is not a preferable imaging method for analysing block excavations as an extended X-ray prospection method. A better analysis of partly exposed samples might be possible by reducing the sensitivity of the setup. This could, for example, be simply realised by increasing the distance from the sample to the phase-grating [38]. A smaller sensitivity leads to a reduced dark-field signal. The amount of reduction depends on the size and shape of the scattering structures [39]. There might be a chance to suppress the influence of soil particles, partially in order to reveal scattering structures of quite different sizes than the soil particles. Here, further investigations are necessary.

Finally, we employed the directional dark-field imaging method on a bird fibula showing mineralised woven textile structures. We successfully reconstructed sub-pixel information about the main fibre orientation of the fabric remnants. Therefore, the directional imaging method proved feasible as an additional analysis tool for the examination of organic remains on archaeological findings. We found that X-ray dark-field imaging is useful in particular regarding the examination and discrimination of fabric remnants attached to archaeological findings. We propose a further application of this technique in particular on wet organic archaeological findings, where the textiles themselves are preserved. The distinction of different types of (mineralised) fibre-based materials, such as fabric remnants, wood, ivory or horn could be another promising application. Furthermore, the analysis of samples concealing internal structures, which are not visible from the outside, could be feasible by directional X-ray dark-field imaging. For further analysis, X-ray phase-contrast computed tomography should be considered. In this regard, the application of dual-energy methods seems a possible technique for the separation of different materials [40,41] in archaeological findings. Furthermore, the application of X-ray dark-field tomography is worth considering [42]. Here, tensor tomography extends the directional dark-field imaging to a 3D analysis of isotropic and anisotropic scattering contributions [21,22,43]. In conclusion, we found that especially X-ray dark-field imaging enables access to several additional analysis methods for the investigation of archaeological findings regarding material composition and structure, even at sub-pixel sized length scales.

Acknowledgments: This work was carried out with the support of the Karlsruhe Nano Micro Facility (KNMF), a Helmholtz Research Infrastructure at Karlsruhe Institute of Technology (KIT).

Author Contributions: Georg Pelzer and Jens Rieger conceived and designed the experimental setup; Veronika Ludwig, Tracy Niepold and Julia Ziegler performed the measurements; Tracy Niepold performed the archaeological interpretation of the samples; Maria Seifert contributed to analysis tools and to scientific discussion; Veronika Ludwig analysed the data and wrote the paper; Thilo Michel and Gisela Anton contributed to scientific discussion and supervised the project.

Conflicts of Interest: The authors declare no conflict of interest.

References

1. Gasteiger, S. Strategien zum Umgang mit archäologischen Eisenfunden in der bayerischen Bodendenkmalpflege. In *Fundmassen: Innovative Strategien zur Auswertung Frühmittelalterlicher Quellenbestände*; Materialhefte zur Archäologie in Baden-Württemberg 97; Brather, S., Krausse, D., Eds.; Konrad Theiss Verlag: Stuttgart, Germany, 2013; pp. 115–130.
2. Nowak-Böck, B. Erhalt und Erfassung organischer Befunde in der bayerischen Bodendenkmalpflege. In *Fundmassen: Innovative Strategien zur Auswertung Frühmittelalterlicher Quellenbestände*; Materialhefte zur Archäologie in Baden-Württemberg 97; Brather, S., Krausse, D., Eds.; Konrad Theiss Verlag: Stuttgart, Germany, 2013; pp. 131–143.
3. Casali, F. X-ray and neutron digital radiography and computed tomography for cultural heritage. In *Physical Techniques in the Study of Art, Archaeology and Cultural Heritage*; Elsevier: Amsterdam, The Netherlands, 2006; Volume 1, pp. 41–123.
4. Morigi, M.; Casali, F.; Bettuzzi, M.; Brancaccio, R.; d'Errico, V. Application of X-ray computed tomography to cultural heritage diagnostics. *Appl. Phys. A* **2010**, *100*, 653–661.
5. Mantler, M.; Schreiner, M. X-ray fluorescence spectrometry in art and archaeology. *X-ray Spectrom.* **2000**, *29*, 3–17.
6. Pfeiffer, F.; Weitkamp, T.; Bunk, O.; David, C. Phase retrieval and differential phase-contrast imaging with low-brilliance X-ray sources. *Nat. Phys.* **2006**, *2*, 258–261.
7. Weitkamp, T.; Diaz, A.; David, C.; Pfeiffer, F.; Stampanoni, M.; Cloetens, P.; Ziegler, E. X-ray phase imaging with a grating interferometer. *Opt. Express* **2005**, *13*, 6296–6304.
8. Revol, V.; Jerjen, I.; Kottler, C.; Schütz, P.; Kaufmann, R.; Lüthi, T.; Sennhauser, U.; Straumann, U.; Urban, C. Sub-pixel porosity revealed by X-ray scatter dark field imaging. *J. Appl. Phys.* **2011**, *110*, doi:10.1063/1.3624592.
9. Pfeiffer, F.; Bech, M.; Bunk, O.; Kraft, P.; Eikenberry, E.F.; Brönnimann, C.; Grünzweig, C.; David, C. Hard-X-ray dark-field imaging using a grating interferometer. *Nat. Mater.* **2008**, *7*, 134–137.
10. Yashiro, W.; Terui, Y.; Kawabata, K.; Momose, A. On the origin of visibility contrast in X-ray Talbot interferometry. *Opt. Express* **2010**, *18*, 16890–16901.
11. Yashiro, W.; Momose, A. Effects of unresolvable edges in grating-based X-ray differential phase imaging. *Opt. Express* **2015**, *23*, 9233–9251.
12. Jensen, T.H.; Bech, M.; Bunk, O.; Donath, T.; David, C.; Feidenhans'l, R.; Pfeiffer, F. Directional X-ray dark-field imaging. *Phys. Med. Biol.* **2010**, *55*, 3317.
13. Jensen, T.H.; Bech, M.; Zanette, I.; Weitkamp, T.; David, C.; Deyhle, H.; Rutishauser, S.; Reznikova, E.; Mohr, J.; Feidenhans'l, R.; et al. Directional X-ray dark-field imaging of strongly ordered systems. *Phys. Rev. B* **2010**, *82*, 214103.
14. Bayer, F.; Zabler, S.; Brendel, C.; Pelzer, G.; Rieger, J.; Ritter, A.; Weber, T.; Michel, T.; Anton, G. Projection angle dependence in grating-based X-ray dark-field imaging of ordered structures. *Opt. Express* **2013**, *21*, 19922–19933.
15. Potdevin, G.; Malecki, A.; Biernath, T.; Bech, M.; Jensen, T.H.; Feidenhans, R.; Zanette, I.; Weitkamp, T.; Kenntner, J.; Mohr, J.; et al. X-ray vector radiography for bone micro-architecture diagnostics. *Phys. Med. Biol.* **2012**, *57*, 3451.

16. Schaff, F.; Malecki, A.; Potdevin, G.; Eggl, E.; Noël, P.B.; Baum, T.; Garcia, E.G.; Bauer, J.S.; Pfeiffer, F. Correlation of X-ray vector radiography to bone micro-architecture. *Sci. Rep.* **2014**, *4*, 3695.
17. Baum, T.; Eggl, E.; Malecki, A.; Schaff, F.; Potdevin, G.; Gordijenko, O.; Garcia, E.G.; Burgkart, R.; Rummeny, E.J.; Noël, P.B.; et al. X-ray Dark-Field Vector Radiography—A Novel Technique for Osteoporosis Imaging. *J. Comput. Assist. Tomogr.* **2015**, *39*, 286–289.
18. Revol, V.; Kottler, C.; Kaufmann, R.; Neels, A.; Dommann, A. Orientation-selective X-ray dark field imaging of ordered systems. *J. Appl. Phys.* **2012**, *112*, 114903.
19. Revol, V.; Plank, B.; Kaufmann, R.; Kastner, J.; Kottler, C.; Neels, A. Laminate fibre structure characterisation of carbon fibre-reinforced polymers by X-ray scatter dark field imaging with a grating interferometer. *NDT E Int.* **2013**, *58*, 64–71.
20. Prade, F.; Schaff, F.; Senck, S.; Meyer, P.; Mohr, J.; Kastner, J.; Pfeiffer, F. Nondestructive characterization of fiber orientation in short fiber reinforced polymer composites with X-ray vector radiography. *NDT E Int.* **2017**, *86*, 65–72.
21. Malecki, A.; Potdevin, G.; Biernath, T.; Eggl, E.; Willer, K.; Lasser, T.; Maisenbacher, J.; Gibmeier, J.; Wanner, A.; Pfeiffer, F. X-ray tensor tomography. *Europhys. Lett.* **2014**, *105*, 38002.
22. Bayer, F.L.; Hu, S.; Maier, A.; Weber, T.; Anton, G.; Michel, T.; Riess, C.P. Reconstruction of scalar and vectorial components in X-ray dark-field tomography. *Proc. Nat. Acad. Sci. USA* **2014**, *111*, 12699–12704.
23. Vogel, J.; Schaff, F.; Fehringer, A.; Jud, C.; Wieczorek, M.; Pfeiffer, F.; Lasser, T. Constrained X-ray tensor tomography reconstruction. *Opt. Express* **2015**, *23*, 15134–15151.
24. Arrizón, V.; López-Olazagasti, E. Binary phase grating for array generation at 1/16 of Talbot length. *J. Opt. Soc. Am. A* **1995**, *12*, 801–804.
25. Suleski, T.J. Generation of Lohmann images from binary-phase Talbot array illuminators. *Appl. Opt.* **1997**, *36*, 4686–4691.
26. Fitzgerald, R. Phase-sensitive X-ray imaging. *Phys. Today* **2000**, *53*, 23–26.
27. Wang, Z.; Stampanoni, M. Quantitative X-ray radiography using grating interferometry: A feasibility study. *Phys. Med. Biol.* **2013**, *58*, 6815–6826.
28. Scherer, K.; Braig, E.; Ehn, S.; Schock, J.; Wolf, J.; Birnbacher, L.; Chabior, M.; Herzen, J.; Mayr, D.; Grandl, S.; et al. Improved Diagnostics by Assessing the Micromorphology of Breast Calcifications via X-Ray Dark-Field Radiography. *Sci. Rep.* **2016**, *6*, 36991.
29. Nesterets, Y.I. On the origins of decoherence and extinction contrast in phase-contrast imaging. *Opt. Commun.* **2008**, *281*, 533–542.
30. Revol, V.; Kottler, C.; Kaufmann, R.; Cardot, F.; Niedermann, P.; Jerjen, I.; Lüthi, T.; Straumann, U.; Sennhauser, U.; Urban, C. Sensing small angle scattering with an X-ray grating interferometer. In Proceedings of the 2010 IEEE on Nuclear Science Symposium Conference Record (NSS/MIC), Knoxville, TN, USA, 30 October–6 November 2010; pp. 892–895.
31. Rieger, J.; Meyer, P.; Horn, F.; Pelzer, G.; Michel, T.; Mohr, J.; Anton, G. Optimization procedure for a Talbot-Lau X-ray phase-contrast imaging system. *J. Instrum.* **2017**, *12*, P04018.
32. Meyer, P.; Schulz, J. Chapter 16—Deep X-ray Lithography. In *Micromanufacturing Engineering and Technology*, 2nd ed.; Qin, Y., Ed.; Micro and Nano Technologies, William Andrew Publishing: Boston, MA, USA, 2015; pp. 365–391.
33. Ritter, A.; Bartl, P.; Bayer, F.; Gödel, K.C.; Haas, W.; Michel, T.; Pelzer, G.; Rieger, J.; Weber, T.; Zang, A.; et al. Simulation framework for coherent and incoherent X-ray imaging and its application in Talbot-Lau dark-field imaging. *Opt. Express* **2014**, *22*, 23276–23289.
34. Siemens.com. Simulation of X-ray Spectra. 2016. Available online: www.siemens.com/x-ray-spectra (accessed on 28 March 2018).
35. Yashiro, W.; Vagovič, P.; Momose, A. Effect of beam hardening on a visibility-contrast image obtained by X-ray grating interferometry. *Opt. Express* **2015**, *23*, 23462–23471.
36. Pelzer, G.; Anton, G.; Horn, F.; Rieger, J.; Ritter, A.; Wandner, J.; Weber, T.; Michel, T. A beam hardening and dispersion correction for X-ray dark-field radiography. *Med. Phys.* **2016**, *43*, 2774–2779.
37. Hubbell, J.H.; Seltzer, S.M. National Institute of Standards and Technology, X-Ray Mass Attenuation Coefficients. 2017. Available online: https://physics.nist.gov/PhysRefData/XrayMassCoef/ComTab/concrete.html (accessed on 28 February 2018).

J. Imaging **2018**, *4*, 58

38. Donath, T.; Chabior, M.; Pfeiffer, F.; Bunk, O.; Reznikova, E.; Mohr, J.; Hempel, E.; Popescu, S.; Hoheisel, M.; Schuster, M.; et al. Inverse geometry for grating-based X-ray phase-contrast imaging. *J. Appl. Phys.* **2009**, *106*, 054703.
39. Strobl, M.; Betz, B.; Harti, R.; Hilger, A.; Kardjilov, N.; Manke, I.; Gruenzweig, C. Wavelength-dispersive dark-field contrast: Micrometre structure resolution in neutron imaging with gratings. *J. Appl. Crystallogr.* **2016**, *49*, 569–573.
40. Johnson, T.R. Dual-energy CT: General principles. *Am. J. Roentgenol.* **2012**, *199*, S3–S8.
41. Kottler, C.; Revol, V.; Kaufmann, R.; Urban, C. Dual energy phase contrast X-ray imaging with Talbot-Lau interferometer. *J. Appl. Phys.* **2010**, *108*, 114906.
42. Bech, M.; Bunk, O.; Donath, T.; Feidenhans, R.; David, C.; Pfeiffer, F. Quantitative X-ray dark-field computed tomography. *Phys. Med. Biol.* **2010**, *55*, 5529.
43. Sharma, Y.; Wieczorek, M.; Schaff, F.; Seyyedi, S.; Prade, F.; Pfeiffer, F.; Lasser, T. Six dimensional X-ray Tensor Tomography with a compact laboratory setup. *Appl. Phys. Lett.* **2016**, *109*, 134102.

Journal of
Imaging

MDPI

Review

State of the Art of X-ray Speckle-Based Phase-Contrast and Dark-Field Imaging

Marie-Christine Zdora [1,2]

[1] Diamond Light Source, Harwell Science and Innovation Campus, Didcot, Oxfordshire OX11 0DE, UK; marie-christine.zdora@diamond.ac.uk

[2] Department of Physics & Astronomy, University College London, London WC1E 6BT, UK

Received: 24 March 2018; Accepted: 17 April 2018; Published: 25 April 2018

Abstract: In the past few years, X-ray phase-contrast and dark-field imaging have evolved to be invaluable tools for non-destructive sample visualisation, delivering information inaccessible by conventional absorption imaging. X-ray phase-sensing techniques are furthermore increasingly used for at-wavelength metrology and optics characterisation. One of the latest additions to the group of differential phase-contrast methods is the X-ray speckle-based technique. It has drawn significant attention due to its simple and flexible experimental arrangement, cost-effectiveness and multimodal character, amongst others. Since its first demonstration at highly brilliant synchrotron sources, the method has seen rapid development, including the translation to polychromatic laboratory sources and extension to higher-energy X-rays. Recently, different advanced acquisition schemes have been proposed to tackle some of the main limitations of previous implementations. Current applications of the speckle-based method range from optics characterisation and wavefront measurement to biomedical imaging and materials science. This review provides an overview of the state of the art of the X-ray speckle-based technique. Its basic principles and different experimental implementations as well as the the latest advances and applications are illustrated. In the end, an outlook for anticipated future developments of this promising technique is given.

Keywords: X-ray speckle-based imaging; X-ray near-field speckle; X-ray phase-contrast imaging; X-ray dark-field imaging; X-ray multimodal imaging; X-ray phase tomography; X-ray wavefront sensing; metrology; optics characterisation

1. Introduction

The first large-scale applications of X-ray imaging can be found in the medical field soon after the discovery of X-rays by Röntgen [1,2]. Radiography [3–6] and a few decades later computed tomography (CT) [7–9] quickly became routine methods in clinics.

In the mid-20th century, the discovery of X-ray synchrotron radiation and the later construction of dedicated synchrotron radiation facilities [10,11] allowed access to X-rays much more powerful than produced by conventional tube sources. The increasing availability of synchrotrons and in particular the development of high-brilliance third-generation synchrotron sources [12] at the end of the 20th century significantly pushed the advances in X-ray science. Soon, imaging with unprecedented resolution and image quality was achieved.

It was with the progresses in X-ray sources and optics in the late 20th century that the great potential of X-ray phase-contrast imaging was first realised. The principle of phase contrast, discovered for visible light in the 1930s [13], had been successfully translated to the X-ray regime in 1965 [14], but received increased attention only with the advent of third-generation synchrotrons providing wider access to highly brilliant, coherent and monochromatic X-rays. Since then, X-ray phase-contrast imaging has seen numerous developments and found a large range of applications, e.g., for biomedical, pre-clinical and clinical imaging [15–25], materials science [26–29], as well as metrology and wavefront sensing [30–40], amongst others.

The main advantage of phase-contrast imaging in the hard X-ray regime over conventional absorption imaging is the significantly higher sensitivity to small density differences [15]. Samples with a low atomic number generally show only little contrast in X-ray attenuation images, while the induced X-ray phase shift can be several orders of magnitude higher, leading to greatly improved signal contrast. However, the practical implementation of X-ray phase-contrast imaging is not straightforward. X-ray detectors can only measure the intensity and not the phase of the X-ray wavefront, so ways had to be found to translate the phase shift into detectable intensity differences. In the past decades, a number of X-ray phase-sensitive imaging methods have been developed [17], starting from the first demonstration by Bonse and Hart using a crystal interferometer [14,41] and followed by analyser-based (diffraction-enhanced) [42–46] and propagation-based (in-line) phase-contrast imaging [47–52], as well as Talbot(–Lau) grating interferometry [53–57] and the edge-illumination and coded-aperture approaches [58–62]. A detailed description of all of these methods would go beyond the scope of this article, but the interested reader is referred to the cited literature.

In this review, we focus on the most recent addition to the group of phase-sensitive imaging methods, namely X-ray speckle-based imaging [63–65]. X-ray speckle-based imaging [65], as well as grating interferometry [66] and analyser-based imaging [45], allow one to reconstruct, in addition to the phase-contrast signal, also the so-called dark-field image, which is a measure of small-angle scattering from features in the sample that cannot be resolved directly [67,68].The dark-field signal can deliver valuable complementary information about the specimen and has recently been used increasingly for medical applications [69–77] and materials science [78–84].

Since it was first proposed a few years ago, X-ray speckle-based phase-contrast and dark-field imaging has drawn significant attention due to its simple, robust and flexible experimental arrangement, cost-effectiveness and relatively low spatial and temporal coherence requirements. These properties also led to the swift translation of the technique to polychromatic laboratory sources [85–87] and its extension from two-dimensional (2D) projection imaging to three-dimensional (3D) tomography implementation [88,89]. Furthermore, it has been shown that, in addition to its great potential for phase-contrast and dark-field imaging for the investigation and visualisation of specimens, X-ray near-field speckle can be employed in the field of metrology for highly precise and accurate X-ray optics characterisation, beam phase sensing and beam coherence measurements [65,90–93].

Despite being a relatively recent approach, the rapid development and increasing interest in the X-ray near-field speckle method promise a widespread implementation and expanding range of applications in the near future.

This review provides an overview of the principles and state of the art of the X-ray speckle-based imaging and metrology technique. Starting from the basic concept of X-ray near-field speckle, the different experimental implementations with their advantages, limitations and challenges are discussed, followed by a more detailed description of the proposed dark-field reconstruction approaches. Subsequently, further progress such as the translation to laboratories and the extension to tomographic imaging is illustrated. In the end, recent applications of the technique are shown, and a summary and outlook for anticipated future developments are given.

2. Basic Principles of X-ray Speckle-Based Multimodal Imaging

2.1. X-ray Speckle as a Wavefront Marker

A speckle pattern is created when (partially) coherent light impinges on an object consisting of randomly distributed scatterers. The phenomenon of speckle has been explored extensively for laser light [94,95]. Laser speckle is on the one hand often an undesired effect, e.g., for laser-based displays [96–98] and in coherent optical imaging [99,100]. On the other hand, it has found many applications such as speckle imaging in astronomy [101–103], electronic speckle pattern interferometry for stress, strain and vibration measurements of rough surfaces [104–108] and dynamic speckle for the investigation of biological processes [109–114]. Even the use of laser speckle for eye testing has been demonstrated [115].

The phenomenon of speckle exists in the far- as well as the near-field regime. However, it is important to note that the properties of these two types of speckle patterns are fundamentally different. While far-field speckles are linked to the illuminating beam, its dimensions and wavelength, it has been shown that in the near field the properties of the speckles are closely related to the scattering features themselves, and the speckle size is independent of the propagation distance and the energy of the beam [116–122].

Just ten years ago, it was demonstrated that the concept of near-field speckle can be directly transferred from the optical to the X-ray regime, and the same criteria and properties apply [123].

Here, it should be noted that the first applications of X-ray near-field speckle were reported already a few years earlier, although not explicitly classifying the observed effects as near-field speckle. In [124,125], the authors report on speckle produced by lung tissue when imaging small animals using the propagation-based phase-contrast technique. They show that the arising speckle can enhance the visible appearance of the lungs in the acquired images, and they explain the occurrence of the speckle by multiple refraction of the X-ray beam in the alveoli of the lung and subsequent free-space propagation. In [126], Kim et al. demonstrate the characterisation of blood flow by means of cross-correlation of X-ray near-field speckle created by scattering off the blood cells.

For X-rays, the near-field regime is much more accessible than for visible light due to their short wavelength. Therefore, it is easily possible to record X-ray near-field speckle created by shining an X-ray beam on small randomly scattering structures. Thanks to the special properties of near-field speckle, the speckle size can be controlled by the size of the scattering particles, and distortions of the speckle pattern upon propagation are only determined by the shape of the wavefront [123]. The above properties make X-ray near-field speckle suitable for use as a wavefront marker for sensing the beam phase. This was soon realised, and the first demonstrations of X-ray speckle-based imaging followed [63,64].

The principle of X-ray speckle imaging is simple: an object in the X-ray beam will lead to a distortion of the X-ray wavefront, which can be observed as a modulation of the speckle pattern. The object can be a sample to investigate, but also optical elements in the experimental setup resulting in desired or undesired changes to the beam. By tracking the modulations of the speckle pattern, the differential phase shift of the X-ray wavefront can be obtained, which allows one to determine the refractive properties of the object. Additionally, the transmission and small-angle scattering information of the specimen can be retrieved. Typically, commercially distributed sandpaper, consisting of small silicon carbide grains, or biological filter membranes with μm-sized pores are used as so-called diffusers to produce a speckle pattern. They are available in different grain sizes [127] and with different pore sizes, respectively, which allows controlling the speckle size and visibility [128]. Other materials containing small scattering features such as finely ground sand, glass or similar and even simple cardboard could also be used as diffusers.

The speckle size is an important property of the speckle pattern that has an influence on the quality of the reconstructed multimodal images acquired with the speckle-based technique. As the speckles have an irregular, random shape and a distribution of sizes, only an average speckle size can be estimated. This is done for example by determining the spatial frequency at the maximum of the power spectrum [86,87,123,129] or from a 2D auto-correlation of the speckle pattern [88,128,130–134]. Generally, well-defined, small speckles that can be resolved easily and cover a few pixels in the detector plane are desired. The speckle size sets a limit for the achievable spatial resolution for the single-shot speckle-tracking reconstruction approach (see Section 3.1), whereas it is not as crucial for implementations based on diffuser stepping (see Sections 3.2 and 3.3).

Another important characteristic of the speckle pattern is the speckle visibility or contrast, which is of significant importance for the reconstruction result. A high visibility of the speckle pattern is beneficial for a successful operation of the reconstruction algorithm. The speckle contrast v is commonly defined in one of the following ways:

$$v = \frac{\sigma_I}{\bar{I}}, \tag{1}$$

where σ_I and $\bar{I}$ are the standard deviation and the mean intensity value, respectively, of the speckle pattern in a small region of interest (typically around 150×150 pixels), as, e.g., used in [87,88,133,135,136], or:

$$v = \frac{I_{\max} - I_{\min}}{I_{\max} + I_{\min}}, \tag{2}$$

where $I_{\max}$ and $I_{\min}$ are the maximum and minimum intensity values of the speckle pattern in a region of interest, as, e.g., used in [85,137,138], or

$$v = \frac{I_{\max} - I_{\min}}{2\bar{I}}, \tag{3}$$

as, e.g., used in [86,135].

It should be noted that the above ways to quantify the visibility of the pattern will give different values and should not be directly compared to each other. Generally Equations (2) and (3) will result in a higher v. This is due to the fact that in these definitions, which use the maximum and minimum intensity values, extreme outliers strongly influence and artificially increase the measured visibility. Furthermore, the impact of outliers makes the quantification of visibility using Equations (2) and (3) somewhat unstable as the measured visibility values will change for different realisations of the speckle map from the same setup and even for different regions of a single speckle image. Equation (1) on the other hand will give a more reliable and stable result as outliers have less effect on the visibility calculation.

2.2. Differential Phase, Transmission and Dark-Field Signals

As mentioned above, near-field speckle can be used to obtain information about the phase shift of X-rays in an object. In addition to the phase-contrast signal, the method also allows the reconstruction of the sample's X-ray transmission and small-angle scattering properties, which can carry valuable complementary information. The principles of image formation for the different signals are outlined in this section.

To keep the explanations and formulas simple for the sake of clarity, in the following we only consider a parallel beam as it is given at synchrotron X-ray sources to a good approximation. However, the concepts discussed here can easily be applied to diverging sources, as mostly encountered in the laboratory [85,87] or for microscopy applications with a magnifying geometry implemented, e.g., with a Fresnel zone plate [139]. The concepts and reconstruction approaches presented here still hold in these cases, but one needs to take into account the magnification of the speckle pattern and sample. A magnifying geometry allows one to significantly increase the spatial resolution by decreasing the effective pixel size in the sample plane, while maintaining a high angular sensitivity that can be influenced by the distances between the (secondary) source, sample and diffuser.

The basic setup for a speckle imaging experiment is shown in Figure 1a. An X-ray beam impinges on a diffuser, e.g., a piece of sandpaper, producing a random speckle pattern, the reference interference pattern, in the detector plane. When a sample is inserted into the beam, the speckle pattern is modulated by the presence of the sample, and this sample interference pattern is recorded by the detector. The modulation appears in three ways as illustrated in Figure 1b: The speckles are displaced in the horizontal x and vertical y directions by a vector $\mathbf{u} = (u_x, u_y)$ due to refraction in the specimen; the mean intensity changes due to absorption; and the visibility of the pattern, i.e., the amplitude after taking into account the absorption, is reduced due to small-angle scattering from unresolved features. From these effects, the refraction angle $\alpha = (\alpha_x, \alpha_y)$, which is related to the differential phase shift, the transmission T (or absorption $A = 1 - T$) and the dark-field signal D, respectively, can be retrieved in a quantitative manner. The reconstruction is performed in real space and pixel-by-pixel using different analysis methods, depending on the experimental implementation; see Section 3.

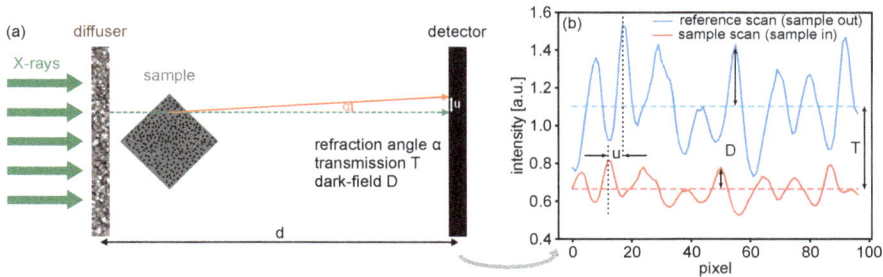

Figure 1. (a) Schematic of an X-ray speckle imaging experiment. X-rays impinge on a diffuser, creating a random speckle pattern in the detector plane. When an absorbing, phase-shifting and scattering sample is placed in the beam, the reference pattern is modulated in intensity, position and visibility. **(b)** Line plot through a few pixels of a reference (blue) and corresponding sample (red) speckle pattern visualising the drop in intensity (dashed horizontal lines) due to absorption $A = (1 - T)$, the displacement u due to refraction of the X-rays by the angle α and the reduction in amplitude (after transmission correction) due to small-angle scattering D.

The displacement of the speckle pattern when a phase-shifting sample is inserted into the beam is visualised in Figure 2 for a phantom sample consisting of a silicon sphere on a wooden toothpick. When looking at a region of interest in the top part of the sphere, it can be observed that the speckles are shifted by several pixels, as can be seen from comparing Figure 2b,c, while outside the sphere no displacement is detected.

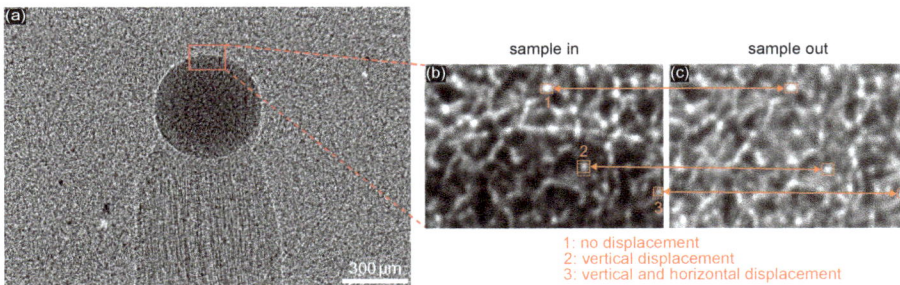

Figure 2. (a) Speckle pattern created by sheets of sandpaper with a sample (silicon sphere of 480 μm diameter glued onto a wooden toothpick) in the beam. Region of interest in **(b)** the sample and **(c)** the reference interference pattern. A displacement of the speckles in (b) with respect to (c) can be observed in the sphere (2, 3), but not in the air background region (1). The dashed orange boxes in **(b)** indicate the corresponding position of the marked speckles in the reference image.

It should be noted that the order of the diffuser and the sample is not crucial for a setup with a parallel beam in combination with the commonly employed experimental implementations used for imaging applications, which are based on comparing reference and sample interference images. However, for the case of diverging beams, the magnification of speckle and sample should be carefully considered. Furthermore, for acquisition schemes that operate without the use of a reference speckle pattern, as often used, e.g., for the characterisation of X-ray mirrors (see Section 7.1), the two configurations give different information (see Section 3.2.3).

The reconstructed transmission signal is obtained from the ratio of the measured local intensities in the sample and reference interference patterns and is similar to the image that would have been obtained without any optical elements and just the sample in the beam. It shows the absorption of the X-rays in the sample, but typically also contains contributions from edge enhancement effects that arise upon propagation in the near-field regime. In this sense, it is not a pure transmission or absorption signal. However, the expression "transmission signal" is still used throughout this article as it is the term commonly found for this contrast modality in the literature on X-ray speckle-based imaging.

The dark-field is given as the local ratio of sample and reference visibilities, i.e., it measures the local reduction in amplitude of the speckle pattern after correcting for the transmission. The dark-field signal can be obtained with different analysis approaches, as discussed in Section 4.

The phase information about the sample is delivered as the differential phase signal, measuring the first derivative of the phase shift induced by the specimen. The direct output from the reconstruction is the displacement **u** of the speckle pattern, which can be converted into the refraction angle α of the X-rays. With the speckle method, the displacement can be measured in the horizontal and vertical directions separately from a single dataset, and the refraction angle $\alpha = (\alpha_x, \alpha_y)$ is obtained. The refraction angle is directly related to the differential phase shift $(\partial\Phi/\partial x, \partial\Phi/\partial y)$:

$$\frac{\partial\Phi}{\partial x} = \frac{2\pi}{\lambda}\alpha_x$$
$$\frac{\partial\Phi}{\partial y} = \frac{2\pi}{\lambda}\alpha_y,$$

(4)

where λ is the X-ray wavelength. The information from the differential phase signals in the two orthogonal directions can then be combined via phase integration to obtain the phase shift Φ of the wavefront. This can be done with various methods, e.g., Fourier-based approaches [64,139–141], two-dimensional numerical integration using least-squares minimisation [91,142] or matrix inversion [135].

2.3. Practical Experimental Considerations

For a practical implementation of the speckle-based technique, there are a few main points to consider. They can be found throughout this article, but are summarised in this section.

The method relies on the use of the near-field speckle pattern as a wavefront marker. For this purpose, speckles should be fully resolved with the detector system to achieve a well-defined speckle pattern of good contrast, yet not too large, in particular for the single-shot implementation of speckle imaging (see Section 3.1), as discussed in Section 2.1. Hence, one should aim for a speckle size in the range of a few times the effective pixel size. As mentioned in Section 2.1, a high speckle visibility improves the quality of the reconstructed images.

The size and visibility of the speckles can be controlled by the type of diffuser used to create the interference pattern. As discussed above, in the near field the speckle dimensions are directly related to the size of the scattering features. Commonly used diffuser materials are commercial abrasive paper and biological filter membranes. The size of the silicon carbide grains of the sandpaper or the membrane pores can be chosen depending on the desired speckle size. Generally, any object containing small scattering particles can be used, and the exploration of further suitable diffuser materials is anticipated.

Regarding the X-ray source, the requirements imposed by the speckle-based technique are moderate. As demonstrated in Section 5, X-ray speckle-based imaging can be performed at polychromatic laboratory sources [85], and the demands on the temporal coherence are low [64,133]. For the creation of speckle, which is based on interference effects, a certain degree of spatial coherence of the X-ray source is required, and the coherence length of the X-ray beam at the diffuser should be larger than the size of the scatterers for an optimum speckle pattern. Reduced spatial coherence leads to a blurring of the interference pattern, which will deteriorate the reconstructed images. It has been demonstrated that microfocus laboratory sources provide sufficient spatial coherence for successful implementation of X-ray speckle-based imaging [85]. Alternatively, it has been proposed to use a random pattern created by absorption rather than interference effects to track the beam wavefront under conditions of low spatial coherence [89,136]; see Section 5.

Another point to consider are setup instabilities during image acquisition. Instabilities in the diffuser or beam position, caused, e.g., by mechanical or thermal instabilities of the diffuser mounting or of optical elements in the beam, lead to a displacement of the speckle pattern in the detector plane. This displacement can be corrected for if the sample is smaller than the field of view by realigning the sample and reference images to the same positions. For the speckle-scanning methods (see Section 3.2), this can give rise to artefacts due to the change in effective step size.

Speckle-based phase-contrast imaging is quite robust against intensity fluctuations of the X-ray beam. As long as the speckle visibility is sufficient, intensity changes between reference and sample scans do not have an impact on the measured refraction signal. They can be observed in the transmission and dark-field images, but corrections can also be performed.

2.4. Related Techniques

The principle of observing the modulations of an X-ray reference pattern to get information about an object is not new, and other established X-ray imaging methods rely on the same phenomenon. For example, X-ray grating interferometry uses a 1D [53–55] or 2D [143,144] phase grating to produce a periodic reference interference pattern in the detector plane. Typically, the fine pattern cannot be resolved directly, and a second so-called analyser grating, placed in front of the detector, is used in combination with a phase-stepping or moiré fringe acquisition approach to translate the pattern into measurable intensity variations in the detector pixels. It has been shown that X-ray grating interferometry in a phase-stepping implementation can in fact be described as a special case of the speckle scanning mode in Section 3.2 [65]. A direct experimental comparison of speckle-based imaging and grating interferometry can be found in [145]. The use of a 1D [146] or 2D [147] transmission grid pattern for the analysis of the sample-induced changes to the reference pattern instead of a phase grating has also been reported for single-shot imaging with the spatial-harmonic technique. The reconstruction processes of the described grating-based methods all exploit the periodic nature of the interference pattern and are based on Fourier transformation.

An alternative reconstruction approach is the analysis by cross-correlation in real space. This was demonstrated with a 1D periodic phase grating [148], a 2D attenuation grid [149] and a 2D phase grid [150]. In contrast to the above mentioned grid method by Wen et al. [147], this approach enables quantitative single-shot imaging of objects with features similar or smaller than the grid pitch as well as objects larger than the field of view. Furthermore, in principle, a periodic structure is not necessary for this approach, and the grid pattern can be replaced by a random interference pattern. In this sense, the single-grid method can be seen as a precursor to X-ray single-shot speckle tracking, discussed in Section 3.1, which uses the same analysis concept.

Compared to the approaches using gratings and grids, speckle-based imaging does not suffer from phase-wrapping effects that can occur for periodic reference patterns. However, artefacts can still arise at the edges of the sample or sample features due to the strong distortions of the speckle pattern in this region, particularly caused by the mixing of the speckle pattern and edge enhancement fringes. This can be reduced by a shorter propagation distance, which, however, also affects the sensitivity of the measurement; see Section 3.4. Furthermore, methods have been investigated to mitigate these artefacts, e.g., by considering the effect of the second derivative of the wavefront [91] or attempting to eliminate the edge effect from the image before reconstruction [151].

As with the single-grid real-space method, large samples with periodic structures of any pitch can be imaged with the speckle-based technique using an easily implemented experimental setup that does not require precise alignment as would be necessary for the two gratings in X-ray grating interferometry. Moreover, the setup for speckle imaging is flexible and the propagation distance can be chosen in the near field without any restrictions that are imposed by the fractional Talbot distances [152] for grating-based imaging with a phase grating.

Furthermore, the use of commercially available sandpaper is very cost-effective and enables accessing the refraction information in the horizontal as well as the vertical directions without the need for elaborate fabrication of 2D structures with small, high-precision period.

It should be noted that one needs to be careful for samples that produce speckle themselves at a similar size to the speckle of the reference pattern. This can lead to artefacts in the reconstruction as the algorithm might not be able to distinguish between the reference interference pattern and the "sample speckle". The size of the reference speckle should be carefully chosen in these cases.

From a resources point of view, the reconstruction in real space is more computationally expensive than Fourier-based algorithms. Fast processing can, however, be achieved by GPU-based computation.

For many applications, the advantages of the speckle-based technique outweigh its challenges. In particular, some of the limitations of both speckle- and grating-based imaging can be overcome simultaneously by our recently proposed advanced operational approach, which can be applied to both random and periodic reference patterns [134]; see Section 3.3.2.

3. Experimental Implementations

The principle of contrast generation in X-ray speckle-based imaging was outlined in Section 2.2. Several operational modes have been developed to quantify the modulations of the interference pattern. The most suitable mode for a certain application depends on the desired speed of data acquisition, spatial resolution and signal sensitivity.

3.1. Single-Shot X-ray Speckle-Tracking Mode (XST)

The first implementations of X-ray speckle-based imaging were demonstrated in single-shot mode, so-called X-ray speckle-tracking (XST), which only requires one reference image with the diffuser, but without the sample in the beam and one sample image with both the diffuser and the sample in the beam [63,64]. As mentioned in the previous section, this approach can be seen as a generalisation of single-shot 2D grid-based methods [149,150] to a random interference pattern.

A schematic of the setup for this approach is shown in Figure 3a. One single sample image is acquired, as illustrated in Figure 3b, and one reference image without the sample; see Figure 3c. As explained in Section 2.2, the X-ray absorption, refraction and small-angle scattering properties of the specimen lead to local changes of the mean intensity, position and visibility of the speckle pattern. Examples of subsets around a pixel of interest in the sample and reference image, respectively, can be seen in Figure 3d,e. The speckle pattern is shifted in Figure 3d compared to the reference in Figure 3e, and the intensity is reduced due to the presence of the sample.

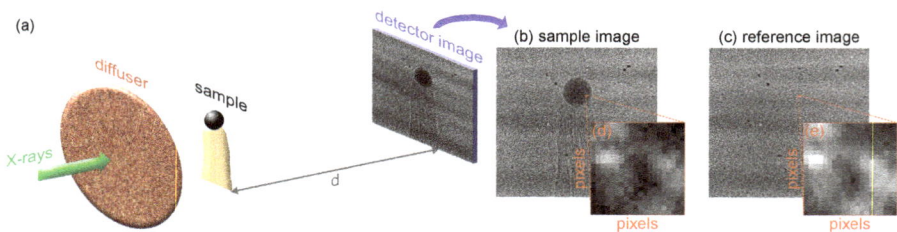

Figure 3. (**a**) Setup for single-shot XST imaging. (**b**) One sample image and (**c**) one reference image (without the sample) are acquired. (**d,e**) A subset window larger than the speckle size is selected around each pixel in the image for reconstruction.

In the XST implementation, the local displacement of the pattern is analysed using a windowed zero-normalised cross-correlation [153] in real space [63–65]. This means that the refraction signal in each pixel of the image is reconstructed by selecting an analysis window around this pixel, as shown in Figure 3d,e, and performing a normalised cross-correlation between the reference and the corresponding sample window.

The cross-correlation coefficient γ between reference and sample windows is given by [63]:

$$\gamma = \sum_{i=-M}^{M} \sum_{j=-M}^{M} \left[\frac{[I_0(x_i, y_j) - \bar{I}_0][I(x_i', y_j') - \bar{I}]}{\Delta I_0 \Delta I} \right]. \tag{5}$$

Here, $I_0(x_i, y_j)$ describes the value in a pixel of the subset of the reference speckle pattern centred at (x_0, y_0) and $I(x_0', y_0')$ the value in a corresponding subset of the sample speckle pattern centred at (x_0', y_0'). The sums run over all pixels in the analysis window of size $2M + 1$. $\bar{I}$ and $\bar{I}_0$ are the mean values and ΔI and ΔI_0 the standard deviations of the sample and reference patterns in the window. If only a rigid translation of the subset is considered, we can write $x_i' = x_i + u_x$ and $y_i' = y_i + u_y$, and the location of the cross-correlation peak $\gamma^{\max}$ corresponds to the local displacement (u_x, u_y) of the speckle pattern in the two orthogonal directions. This can then be converted to a refraction angle signal (α_x, α_y) by geometrical considerations (in small-angle approximation):

$$\begin{aligned} \alpha_x &= \frac{u_x \cdot p_{\text{eff}}}{d} \\ \alpha_y &= \frac{u_y \cdot p_{\text{eff}}}{d}, \end{aligned} \tag{6}$$

where p_{eff} is the effective pixel size in the detector plane and d the propagation distance (The propagation distance d corresponds to the sample-detector distance for the configuration in Figure 3a where the diffuser is placed upstream of the sample, but to the diffuser-detector distance if the diffuser is placed downstream of the sample.). The analysis window slides across the whole image, and refraction, transmission and dark-field signals are obtained locally for each pixel.

The transmission signal can be calculated from the ratio of the mean intensities in the sample and reference windows:

$$T = \bar{I}/\bar{I}_0. \tag{7}$$

The dark-field image is typically retrieved as the ratio of the sample and reference visibilities, which can be quantified for each pixel as the ratio of the standard deviation divided by the mean intensity in the respective sample and reference analysis windows [65]:

$$D = \frac{\Delta I/\bar{I}}{\Delta I_0/\bar{I}_0} = \frac{1}{T}\frac{\Delta I}{\Delta I_0}. \tag{8}$$

It has also been proposed that alternatively the reduction of the cross-correlation peak value can be taken as a measure for the dark-field signal [89]; see Section 4.

A different approach for the image reconstruction of XST data was introduced a bit later [85]. The idea is based on a physical model of the speckle interference pattern in the detector plane that takes into account the modulations of the pattern by the presence of the sample. For a certain pixel (x, y), the sample interference pattern I can be described in terms of the reference interference pattern I_0, but modulated in intensity, amplitude and position by the properties of the sample:

$$I(x, y) = T(x, y) \left[\bar{I}_0 + D(x, y) \left(I_0(x + u_x, y + u_y) - \bar{I}_0 \right) \right]. \tag{9}$$

Here, $\bar{I}_0$ is the mean intensity of the reference pattern and $T(x, y)$ the local transmission through the sample reducing the intensity of the speckle pattern. The amplitude $\left(I_0(x + u_x, y + u_y) - \bar{I}_0 \right)$ of the reference pattern is reduced by the factor $D(x, y)$ corresponding to the local dark-field signal. The refraction in the sample is taken into account by the quantities u_x, u_y, describing the displacement of the interference pattern in the horizontal and vertical directions, respectively. For image reconstruction, a windowed least-squares minimisation between the model in Equation (9) and the measured sample speckle pattern is performed. The minimisation procedure is conducted pixel-by-pixel using the sum over the pixels in an analysis window w around the pixel of interest (x_0, y_0):

$$\mathcal{L} = \sum_{i=-M}^{M} \sum_{j=-M}^{M} w(x_i, y_j) \left\{ I(x_i, y_j) - T(x_i, y_j) \left[\bar{I}_0 + D(x_i, y_j) \left(I_0(x_i + u_x, y_j + u_y) - \bar{I}_0 \right) \right] \right\}^2 \qquad (10)$$

Minimisation of the function $\mathcal{L}$ delivers the multimodal image signals u_x, u_y, T and D. From the speckle displacement (u_x, u_y), the refraction angle (α_x, α_y) can be obtained via Equation (6).

The extent of the analysis window w should be larger than the average speckle size to achieve a good reconstruction result. Different window types can be used from a simple square window with equal weighting for all pixels to Hamming or Tukey (tapered cosine) windows that give less weight to pixels at the edges. The latter can often lead to improved results with reduced artefacts.

Commonly, in the XST analysis approach, as outlined above, only a rigid translation of the speckle pattern is considered, and higher-order modulations of the sample subset compared to the reference subset are neglected. However, it has been shown that considering the distortions of the analysis subset can improve the robustness and accuracy of the reconstruction algorithm and furthermore delivers additional information, e.g., on the local curvature of the X-ray wavefront [91]. The coefficients of the higher-order distortions can be obtained from a minimisation approach after determining the rigid translation of the subset. Consideration of higher-order subset distortions can be beneficial, e.g., for analysing focussing samples such as X-ray refractive lenses. The information from higher-order distortions could in this case help to reduce artefacts arising from the demagnification of the reference pattern in the lens.

The main advantage of the XST implementation is the fast image acquisition, which makes it suitable for dynamic imaging and in-vivo studies. It was demonstrated that a successful reconstruction can be achieved from a single image with sub-μs exposure time at a synchrotron source [128]. Furthermore, XST does not require any special equipment, such as high-accuracy, high-precision scanning stages that are needed for the speckle-scanning method discussed in the next section. As it is essential that the position of the diffuser is identical for the reference and the sample scan, some stability of the setup is required. However, a slight displacement of the speckle pattern caused by drift or movement of the diffuser or beam instabilities can be corrected for by realigning the reference and sample images in the empty space background, e.g., via cross-correlation, as discussed in Section 2.3.

The main drawback of the single-shot approach is the limited spatial resolution that is given by twice the full width at half maximum (FWHM) of the size of the analysis window, which needs to be larger than the speckle size. The ultimate limit for the resolution of this operational mode is the speckle size.

3.2. X-ray Speckle-Scanning Modes (XSS)

For applications where high resolution is more important than image acquisition speed, the speckle-scanning (XSS) mode, also called speckle-stepping mode, is more suitable. It was proposed shortly after the single-shot approach and can be considered as a generalised version of X-ray grating interferometry in phase-stepping mode [65]. However, the analysis of speckle scanning data is performed in real space, as opposed to the Fourier analysis for X-ray grating interferometry. The speckle-scanning mode has been demonstrated in two experimental ways [138]: 2D and 1D scanning, which are described in the following.

3.2.1. 2D Scanning (2D XSS)

The first speckle-stepping implementation was reported for scanning of the diffuser in both the horizontal and the vertical direction in small equidistant steps [65], as illustrated in Figure 4a. This way, a signal is recorded at each diffuser position, with and without the sample in the beam; see Figure 4b,c.

A sample and a reference 2D array, which contain the intensities at each diffuser step with and without the sample in the beam, respectively, are obtained for each pixel in the detector plane. Examples of these arrays for one pixel are shown in Figure 4d,e. The reconstruction can then be performed pixel-wise and effectively in the sample plane (The reconstruction is effectively performed in the sample plane if the sample is placed downstream of the diffuser, but in the diffuser plane if the sample is placed upstream of the diffuser.) by zero-normalised cross-correlation (see Equation (5)) of these sample and reference arrays. The retrieval of the three complementary image signals—transmission, refraction and dark field—is conducted analogous to the single-shot case, but with the analysis arrays built from the signals at different diffuser positions rather than different pixels of an analysis window. The displacement (u_x, u_y) of the speckle pattern between reference and sample arrays is now given in units of diffuser steps and can be converted to a refraction angle signal in the horizontal and vertical direction separately:

$$\alpha_x = \frac{u_x \cdot s}{d}$$
$$\alpha_y = \frac{u_y \cdot s}{d},$$

(11)

where s is the diffuser step size (To be precise: s is the diffuser step size in the sample plane in the case that the sample is placed downstream of the diffuser, but the step size in the diffuser plane in the case that the sample is placed upstream of the diffuser.) and d the propagation distance.

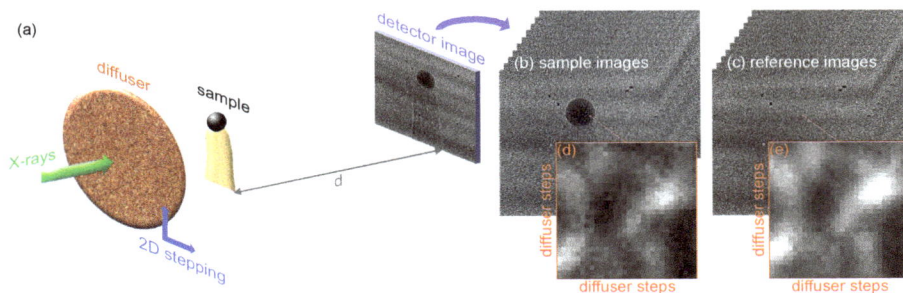

Figure 4. (**a**) Setup for 2D XSS imaging. The diffuser is stepped in two directions in small equidistant steps on a regular grid. (**b**) A sample image and (**c**) a reference image are acquired at each of the several hundred diffuser positions. (**d**,**e**) For each pixel, a 2D array with the signal at each diffuser position is obtained, enabling a high-resolution pixel-wise reconstruction.

The sensitivity of the refraction angle measurement critically depends on the diffuser step size (see Section 3.4), and typically small steps in the range of the pixel size or smaller are chosen. For step sizes much smaller than the effective pixel size, it is important to ensure that the intensity variation between subsequent steps is sufficient. For a certain experimental arrangement, this sets the limit of the achievable angular sensitivity. Typically, the diffuser is scanned on a grid of several tens of steps across in each direction, adding up to a total number of hundreds of frames for the reconstruction of one image, which makes this approach unsuitable for fast imaging applications. Furthermore, due to the small regular step sizes, XSS requires delicate and costly high-accuracy, high-precision scanning stages, which should be aligned carefully with the beam direction to ensure equal step sizes in both directions.

Compared to the XST approach, XSS is significantly more sensitive to instabilities of the setup. The technique requires the speckle pattern to be shifted by a known constant step. Deviations from the desired position of the speckle pattern, caused by instabilities of the beam or setup (see Section 2.3), cannot be corrected for as this would alter the effective step size.

In contrast to the XST approach, however, where several pixels in an analysis window contribute to the signal reconstruction of one pixel, the stepping mode allows a real pixel-wise analysis. This enables a much higher resolution down to the pixel size, which is the main advantage of the XSS technique. In practice, the point-spread function of the detector and other factors might deteriorate the resolution.

3.2.2. 1D Scanning (1D XSS)

To reduce the number of acquired images, it was proposed that two orthogonal 1D scans could be used instead of a full 2D grid scan in cases of small speckle displacement, i.e., for short propagation distances or moderately phase-shifting samples [65].

This was simplified further by taking only one single 1D scan to obtain the 2D refraction information [87]. In this mode, here called 1D XSS, the diffuser is stepped only in one direction—horizontally or vertically—in equidistant steps that are much smaller than the average speckle size and in the order of the pixel size. This is done with and without the sample in the beam; see Figure 5b,c for scanning in the horizontal direction. To be able to track the 2D speckles, a few nearby pixels are selected in the orthogonal direction that is not scanned. For each pixel to be reconstructed, one gets a signal at each diffuser step and takes a 1D window of a few pixels in the other direction, giving a 2D array per pixel. A cross-correlation is now performed between the sample and the reference arrays constructed this way. An example of the signal for one pixel is shown with and without the sample in Figure 5d,e, respectively. Typically several tens of steps are taken in one direction, and only a few pixels are selected in the orthogonal direction. Effectively, the 1D XSS approach can be considered a hybrid between the 2D XSS and the XST cases.

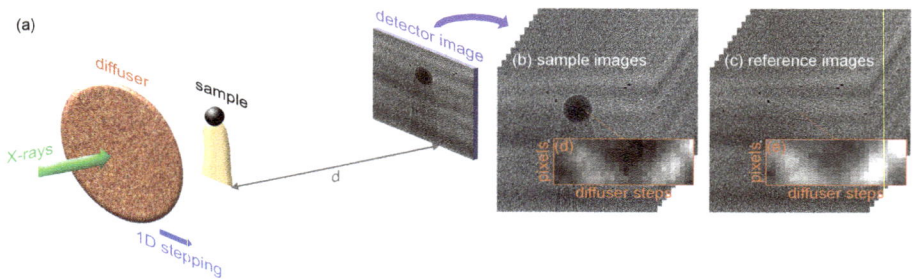

Figure 5. (**a**) Setup for 1D XSS imaging. The diffuser is stepped in only one direction in small equidistant steps. (**b**) A sample image and (**c**) a reference image are acquired at each of the several tens of diffuser positions. (**d,e**) For each pixel, a 2D array is built from the signal at each diffuser position in the pixel of interest and a few surrounding pixels in the direction that is not scanned.

The reconstruction of the transmission and dark-field signals is performed the same way as in the 2D XSS and XST implementations by looking at the local changes in mean intensity and visibility within the analysis arrays for each pixel. The location of the cross-correlation peak gives the displacement (u_x, u_y) of the speckle pattern in the two directions. However, the two axes of the analysis array are not the same, and the displacement is given in units of diffuser steps in the scanning direction and in units of effective pixel size in the orthogonal direction. For horizontal scanning, the conversion from the measured displacement to refraction angle signal is hence given by:

$$\alpha_x = \frac{u_x \cdot s}{d}$$
$$\alpha_y = \frac{u_y \cdot p_{\text{eff}}}{d}, \tag{12}$$

where d is the propagation distance, s the diffuser step size and p_{eff} the effective pixel size.

The 1D XSS approach allows a significantly faster image acquisition than 2D XSS. However, still several tens of frames have to be acquired for a successful image reconstruction. For most experimental realisations of 1D XSS found in the literature, a set of 60 diffuser steps was used, and a minimum of 40 steps has been reported [87,136]. The more steps are taken, the smaller the required number of pixels in the other direction and vice versa. This makes the approach more flexible than 2D XSS, while allowing a better spatial resolution than XST. On the other hand, the spatial resolution is reduced compared to 2D XSS as several surrounding pixels contribute to the reconstruction of the signal in one pixel. Furthermore, the sensitivity of the reconstructed images is not the same in the two orthogonal directions [138]. The sensitivity is generally reduced along the non-scanned direction as it is dependent on the effective pixel size, but on the step size for the scanning direction. 1D XSS could be used effectively in cases with a preferred direction of interest, for which the signals of the sample in the other direction are not crucial. The additional refraction information in the orthogonal direction can still be used to improve and reduce artefacts in the integrated phase signal.

3.2.3. Scanning with Self-Correlation Analysis

For imaging purposes, 2D XSS or 1D XSS is typically performed in differential mode relying on the acquisition of reference and sample scans, as explained in the previous sections. However, for metrology applications (see Section 7.1), in particular for the characterisation of strongly focussing optics, often another mode is used, which is sometimes called self-correlation mode [93]. In this implementation, the local curvature, i.e., the second derivative, of the wavefront is measured as opposed to the first derivative obtained from the commonly employed differential mode [65].

For the self-correlation mode, image acquisition is performed by scanning the diffuser following the same schemes as for the common XSS methods, but no reference images are taken. The correlation procedure is then applied to the signals recorded in two nearby pixels during the same image acquisition. The two pixels that are separated in the detector plane by a pixels in the x-direction and b pixels in the y-direction (i.e., by absolute distances ap_{eff} and bp_{eff}, respectively) will see the same signal, but at different times depending on the diffuser step size s. Cross-correlation between the signals in the two pixels gives the delay $(\chi_x, \chi_y) = (u_x s, u_y s)$ for the observation of the same signal in the pixels, where (u_x, u_y) is the position of the maximum of the correlation coefficient. From geometrical considerations, one can approximate the local radius R of the wavefront in the x- and y-directions as follows [65]:

$$R_x = \frac{d \cdot ap_{\text{eff}}}{ap_{\text{eff}} - \chi_x} = \frac{d \cdot ap_{\text{eff}}}{ap_{\text{eff}} - u_x s}$$
$$R_y = \frac{d \cdot bp_{\text{eff}}}{bp_{\text{eff}} - \chi_y} = \frac{d \cdot bp_{\text{eff}}}{bp_{\text{eff}} - u_y s}.$$

(13)

Here, d is the propagation distance and p_{eff} the effective pixel size. For small angles, the local radius R of the wavefront W is directly related to its local curvature or second derivative, which is in turn proportional to the second derivative of the beam phase Φ:

$$\frac{1}{R_x} \approx \frac{\partial^2 W}{\partial x^2} = \frac{\lambda}{2\pi} \frac{\partial^2 \Phi}{\partial x^2}$$
$$\frac{1}{R_y} \approx \frac{\partial^2 W}{\partial y^2} = \frac{\lambda}{2\pi} \frac{\partial^2 \Phi}{\partial y^2},$$

(14)

where λ is the wavelength.

The self-correlation analysis approach can deliver two different kinds of information, depending on the location of the diffuser: It gives the wavefront distortions induced by the object under study, if the diffuser is mounted upstream of the sample, or the wavefront distortions caused by all optics and components in the beam upstream, if the diffuser is mounted downstream of the sample.

The self-correlation analysis is often used for metrology applications, in particular for the characterisation of X-ray mirrors; see Section 7.1. Self-correlation analysis of 1D scanning data can be applied to obtain the 1D slope of mirrors, and in this case, not the signal delay between two different

single pixels, but between two rows (The signal delay between two rows is analysed if the diffuser is
scanned vertically. For the case of scanning the diffuser horizontally, the signal delay between two
columns is determined.) of the detector image is considered. A 2D array is built for each row by
stacking the signals in the row at each diffuser position. The correlation procedure is applied to the
arrays of two neighbouring rows delivering the delay signal that can then be converted to the local
wavefront curvature [90]. The retrieval of the 2D wavefront curvature can be achieved by scanning
the diffuser in two separate orthogonal 1D scans along the vertical and the horizontal directions.
Furthermore, it has been shown that the 2D information on the wavefront curvature can also be
accessed from only a single 1D scan by looking at the signal delay in two neighbouring pixels along the
scanning direction and at the same time noting the displacement of the speckle pattern in the direction
that is not scanned [154,155]. One should be aware that for this approach, the displacement of the
speckles along the non-scanned axis needs to be small, and the sensitivity in this direction is typically
lower than in the scanning direction, similar to the conventional 1D XSS mode. The technique can be
useful to reduce artefacts in the reconstructed 2D slope profile of a mirror arising when only the 1D
speckle displacement is considered.

3.2.4. 2D Scanning with Sparse Sampling

The 1D XSS approach requires significantly less diffuser steps than 2D XSS, which, however,
comes at the cost of a reduced sensitivity in the direction orthogonal to the scanning axis. Moreover,
still several dozens of diffuser positions are needed for a successful reconstruction. Recently, a stepping
scheme has been proposed that uses the concept of 2D XSS, but with a sparse sampling for the sample
scan [156]. The acquisition of the reference patterns without the specimen in the beam is performed
with a 2D raster scan of the diffuser as for classic 2D XSS, while sample images are taken only at every
n-th point of the diffuser scanning grid of the reference scan. The missing sample images are then
obtained via interpolation. It was demonstrated that a coarse scanning grid of only 5×5 steps for the
sample acquisition and subsequent interpolation to 25×25 step arrays is sufficient to obtain images of
good quality comparable to the full 2D XSS data. This significantly reduces the scan time and dose
to the sample. However, a four-fold reduction of the sensitivity has been reported for the 5×5-step
sparse scanning scheme compared to a conventional 25×25-step 2D XSS scan [156].

3.2.5. Analysis of the Scattering Distribution

A conceptually different approach for retrieving the information about the sample from
2D diffuser scanning data is the recovery of the ultrasmall-angle scattering distribution of the
sample [157]. This was inspired by an analogue reconstruction process introduced for 1D and
2D grating interferometry [158,159]. In general, the sample interference pattern can be expressed
as the convolution of the reference signal without the sample in the beam and the optical transfer
function of the specimen, equivalent to the sample scattering distribution. The reconstruction approach
relies on recovering the optical transfer function using iterative methods like the Richardson–Lucy
deconvolution [160,161]. The moments of the scattering distribution carry different information about
the sample [158,159]. The zeroth moment is equivalent to the transmission through the specimen,
while the first order moments can be interpreted as the differential phase signals in the two directions.
The second moments give directional information about the small-angle scattering strength, equivalent
to the common dark-field signal. The third and fourth moments quantify the skewness and kurtosis of
the distribution, respectively.

Although this approach gives a large number of contrast channels, some of which are not
accessible with the other reconstruction methods, and has been shown to have improved angular
sensitivity [158,159], the cumbersome and computationally expensive reconstruction procedure has so
far impeded its wider implementation for speckle-based imaging.

3.3. Acquisition with Random Diffuser Positions

As discussed in the previous sections, there are some crucial limitations of the classic implementations of speckle-based imaging in the single-shot XST and XSS modes. While the XST approach is quite limited in spatial resolution, the XSS modes require a large number of acquired frames. The 1D XSS scheme results in different sensitivities for the horizontal and vertical directions and a reduced resolution. Recently, efforts have been made to develop experimental implementations that provide a trade-off between the advantages and drawbacks of the two classic modes XST and XSS.

Three approaches for this purpose have been proposed, namely the speckle-vector tracking technique (XSVT) [157], the mixed XSVT approaches [135,156] and the unified modulated pattern analysis (UMPA) [134]. They all rely on taking sample and reference scans at several different diffuser positions, as shown in Figure 6a. In contrast to the XSS mode, in the case of the advanced methods, the diffuser positions can be randomly chosen, and step sizes should be significantly larger than the speckle size. This allows the use of less accurate, less costly stepping stages. They, however, still need to be precise and repeatable to ensure that sample and reference images are taken at the same diffuser positions. The number of required steps is much lower than for the XSS case, allowing shorter scan times.

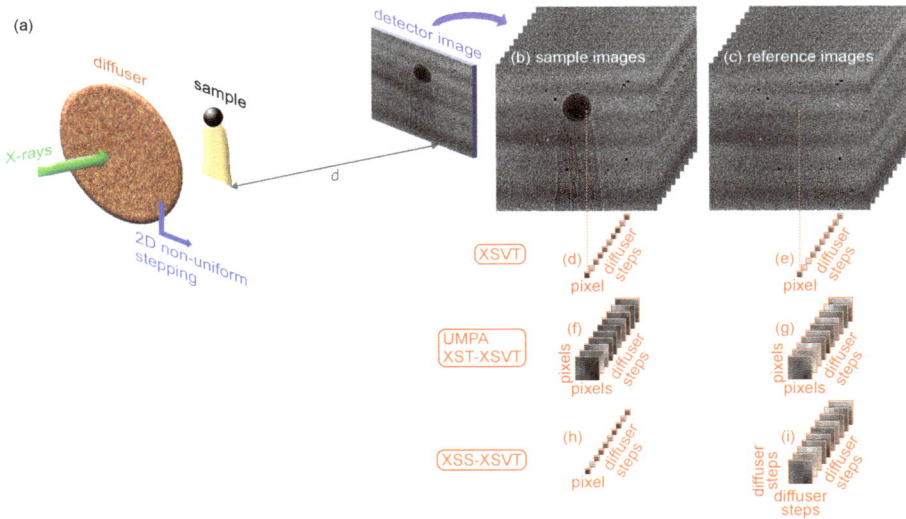

Figure 6. (**a**) Setup for speckle imaging with a random scan pattern of non-equidistant, large steps (XSVT and UMPA). The diffuser is stepped in two directions in a few large random steps. (**b**) A sample image and (**c**) a reference image are acquired at each of the diffuser positions. For XSVT, the analysis is performed for each pixel by comparing the (**d**) sample and (**e**) reference vectors built from the intensity in a single pixel at each diffuser step. For the UMPA and XST-XSVT approaches, a small subset window is chosen around the pixel under consideration in each of the (**f**) sample and (**g**) reference images at the different diffuser positions, allowing one to reduce the number of steps and improving the reconstruction result. For the XSS-XSVT reconstruction, a (**h**) sample vector is built as for the XSVT case, while for (**i**) the reference vector, the diffuser is scanned in small equidistant steps around each of the initial diffuser positions.

3.3.1. X-ray Speckle-Vector Tracking (XSVT) and Mixed XSVT Approaches

The XSVT method considers the signal in each pixel to be a vector made up from the measured intensities at all N diffuser positions. A sample vector $\mathbf{i_r} = (i_{\mathbf{r},1}, \ldots, i_{\mathbf{r},N})$ (see Figure 6d) and a reference vector $\mathbf{o_r} = (o_{r,1}, \ldots, o_{r,N})$ (see Figure 6e) can be created for each pixel $\mathbf{r} = (x,y)$. For the reconstruction of the multimodal images, a zero-normalised cross-correlation is performed between the reference and sample speckle vectors [157]:

$$\gamma(\mathbf{i_r}, \mathbf{o_{r+h}}) = \frac{\sum_{k=1}^{N}(i_{r,k} - \bar{i}_r)(o_{r+h,k} - \bar{o}_{r+h})}{\sqrt{\sum_{k=1}^{N}(i_{r,k} - \bar{i}_r)^2 \sum_{k=1}^{N}(o_{r+h,k} - \bar{o}_{r+h})^2}}, \tag{15}$$

where $\mathbf{h}$ is a small displacement and $\bar{i}_r$ and $\bar{o}_{r+h}$ are the mean values of the sample and reference vectors, respectively, for the pixel $\mathbf{r}$. The position of the correlation peak $\mathbf{u} = \arg\max_{\mathbf{h}} \gamma(\mathbf{i_r}, \mathbf{o_{r+h}})$ gives the displacement $\mathbf{u} = (u_x, u_y)$ of the speckle pattern due to refraction in the sample, which can then be converted into a refraction angle signal using Equation (6). Here, arg max stands for "arguments of the maxima" and $\arg\max_{\mathbf{h}} \gamma(\mathbf{i_r}, \mathbf{o_{r+h}})$ corresponds to the displacement $\mathbf{h}$ for which $\gamma(\mathbf{i_r}, \mathbf{o_{r+h}})$ reaches its maximum value. The transmission signal of a pixel can be obtained from the ratio of the mean intensities of the sample and reference speckle vectors. The dark-field signal is retrieved from the ratio of the standard deviations of the sample and reference speckle vectors normalised by the transmission.

To allow reducing the number of diffuser steps further and to make the method more flexible, a mixed XST-XSVT approach was proposed [135]. The principle of image reconstruction based on the correlation of speckle vectors is the same. However, for each of the acquired images, at the same time, a small analysis window is chosen around the pixel under consideration (similar to XST), as illustrated in Figure 6f,g. The information from the surrounding pixels in the analysis window at the different diffuser positions contributes to the speckle vector of each pixel and is included in the correlation analysis to obtain the image signals. This is done analogous to Equation (15), but the sum now runs not only over all diffuser positions, but also over all pixels in the window.

Recently, also a mixed XSS-XSVT approach has been proposed [156]. The acquisition of the sample interference patterns follows the normal XSVT scheme, and sample images are taken at several random diffuser positions building up a sample vector for each pixel to be reconstructed; see Figure 6h. For the reference images, the diffuser is additionally scanned in small regular steps around each of the diffuser positions of the sample scan. The recorded signals for all of the positions can be arranged in a reference vector for each pixel; see Figure 6i. The reconstruction of the multimodal images is performed by the correlation of sample and reference vectors. The displacement of the speckles is obtained from the location of the cross-correlation peak, and the refraction angle can be calculated via Equation (11).

3.3.2. Unified Modulated Pattern Analysis (UMPA)

The same acquisition scheme as for the combined XST-XSVT approach is used for the recently proposed UMPA method. Images at a few different random diffuser positions are recorded, and a small analysis window around the pixel of interest is applied, as shown in Figure 6f,g. However, UMPA proposes a different concept for data analysis that is based on a least-squares minimisation between a model and the measurement of the sample interference pattern summed over all diffuser positions [134]. This model was first proposed for the XST mode [85]; see Equation (9). For the UMPA approach, the model in Equation (9) holds for each interference pattern at diffuser position n. In the least-squares minimisation process (see Equation (10)) of the function $\mathcal{L}$, the sum now runs not only over all pixels in the analysis window, but also over all diffuser positions n:

$$\mathcal{L} = \sum_{n} \sum_{i=-M}^{M} \sum_{j=-M}^{M} w(x_i, y_j) \left\{ I_n(x_i, y_j) - T(x_i, y_j) \left[\bar{I}_0 + D(x_i, y_j) \left(I_{0n}(x_i + u_x, y_j + u_y) - \bar{I}_0 \right) \right] \right\}^2. \tag{16}$$

Here, $w(x_i, y_j)$ is the window function, which has typically a much smaller extent than for the XST case; $I_n(x_i, y_j)$ and $I_{0n}(x_i, y_j)$ are the intensities in pixel (x_i, y_j) at step n of the diffuser with and without sample, respectively; and $\bar{I}_0$ the mean intensity of the reference pattern over all diffuser positions. The local speckle displacement (u_x, u_y), transmission T and dark-field signal D are obtained directly from the reconstruction, and the refraction angle can be calculated from the displacement using Equation (6).

For both the XST-XSVT and the UMPA approaches, the use of an analysis window around the pixel to be reconstructed allows one to significantly reduce the number of acquired frames by adding information from the surrounding pixels. The size of the analysis window is typically only a few pixels across, resulting in a moderate reduction in spatial resolution. However, the choice of the number of steps and window size are always coupled. The exact parameter combinations depend on the focus of the specific experiment, in particular on the desired spatial resolution and refraction signal sensitivity. Larger window sizes generally allow the use of fewer diffuser steps, but lead to a reduced spatial resolution, while a larger number of diffuser positions enables high-resolution imaging with a small analysis window at the cost of long acquisition times and a high dose to the sample. Therefore, these approaches can be seen as a trade-off between the XST and XSS modes, and they allow flexible tuning of the reconstruction result. This will aid the straightforward implementation of speckle-based techniques for a wider range of applications with different requirements on scan times, spatial resolution and signal sensitivity, also at laboratory sources.

Furthermore, it has been demonstrated that UMPA can be successfully applied not only to random speckle patterns, but also periodic reference patterns such as the Talbot self-image created by a beam-splitter phase grating [134]. This will facilitate the implementation of flexible and tunable phase-contrast and dark-field imaging at most existing X-ray phase-contrast imaging setups without the need for significant modifications.

3.4. Angular Sensitivity and Spatial Resolution

The two main criteria for assessing the quality of the reconstructed phase-contrast images are the spatial resolution and the angular sensitivity.

The spatial resolution strongly depends on the experimental implementation and processing method. For XST (Section 3.1), it is determined by the size of the subset window chosen in the reconstruction process and is ultimately limited by the speckle size. For 2D XSS (Section 3.2.1), it can go down to the effective detector pixel size as a pixel-wise reconstruction is performed. In practice, the point-spread function of the detector and other factors might deteriorate the resolution. For 1D XSS (Section 3.2.2), the spatial resolution is reduced as a few pixels taken along the axis orthogonal to the scanning direction contribute to the signal formation. For the sparse sampling variation of 2D XSS, a pixel resolution could in principle be realised, but due to the interpolation step used for the sample image, it might be lower. The XSVT as well as the mixed XSS-XSVT approaches (Section 3.3.1) can also achieve a resolution down to the pixel size. On the other hand, the mixed XST-XSVT (Section 3.3.1) and UMPA (Section 3.3.2) approaches show a lower spatial resolution that is determined by the extent of the subset window taken around the pixel of interest. Typically, the window sizes are much smaller than for XST, and hence a higher resolution can be achieved with UMPA and XST-XSVT. The resolution limit can be quantified as twice the FWHM of the window extent [134].

The second property commonly used to evaluate the quality of the reconstructed phase-contrast images is the angular sensitivity, which is a measure of the smallest refraction angle or differential phase shift that can be measured with a certain setup and acquisition scheme. The sensitivity is typically quantified as the standard deviation of the reconstructed refraction angle signal in a small region of interest in the air background without sample. As for the spatial resolution, it also strongly depends on the processing scheme. In general, it is inversely proportional to the propagation distance and dependent on the accuracy of the reconstruction algorithm and the photon noise, amongst other factors. A detailed study on the noise, directly related to the angular sensitivity, in the differential phase signals for XST measurements based on simulations and experimental validation can be found in [162]. For the XST, XSVT, mixed XST-XSVT and UMPA methods that perform the reconstruction in the detector plane, the angular sensitivity is furthermore directly proportional to the effective pixel size

p_{eff}. For the approaches that operate in the sample (diffuser) plane (Depending on the mounting of the diffuser upstream or downstream of the sample, the reconstruction is effectively performed in the sample or the diffuser plane, respectively.), such as 2D XSS and mixed XSS-XSVT, it is proportional to the diffuser step size s in the sample (diffuser) plane instead. This means that these operational modes can achieve a better sensitivity for a given setup (An up to 100-fold improvement of the sensitivity for 2D XSS compared to XST has been reported [135].), as s is typically smaller than p_{eff}. For the 1D XSS analysis, a high sensitivity dependent on the step size can be achieved in the scanning direction, whereas the sensitivity in the other direction that is not scanned is proportional to the pixel size.

Further quantities influencing the angular sensitivity are the number N of diffuser steps for the reconstruction of the image and the extent w of the analysis subset. It can be shown that the angular sensitivity is inversely related to w and $\sqrt{N}$ [134]. This relationship makes the UMPA and XST-XSVT approaches very attractive since the angular sensitivity can be controlled by changing N and w. As mentioned above, the choice of w also determines the spatial resolution of the reconstructed images. Hence, the UMPA and XST-XSVT modes allow flexible tuning of the resolution and sensitivity that can be adjusted to specific experimental requirements. In a practical implementation, also the constraints in scan time and dose, which inherently increase with N, might play a role. The choice of N and w ultimately depends on the focus and the desired outcome of the experiment.

4. Speckle-Based X-ray Dark-Field Imaging Approaches

In the very first demonstrations of X-ray speckle-based imaging [63,64], the focus was solely on the phase-contrast signal. However, it was soon recognised that complementary dark-field information can be obtained simultaneously from a speckle imaging dataset [65]. Although the capabilities of the dark-field image have not yet been extensively exploited for X-ray speckle-based imaging applications, there is great potential in particular for dark-field tomography (see Section 6) for medical and materials science applications.

The dark-field signal gives information about the small-angle scattering in the sample [66,68]. For speckle imaging, it is related to the loss in visibility of the speckle pattern caused by a decrease in the coherence of the X-rays after undergoing scattering in the specimen. Different models have been developed to measure the dark-field signal from the acquired speckle data. The first proposed method is analogous to the treatment in X-ray grating interferometry, where the dark-field is defined as the ratio of the amplitudes of the sample and reference phase-stepping curves normalised by the transmission [66]. For speckle-based imaging, the same concept can be used, and the standard deviation of the interference pattern can be taken as a measure for the amplitude. The equivalent description of the dark-field signal D for speckle imaging is then given by the ratio of the sample and reference standard deviations, σ_{sam} and σ_{ref}, normalised by the transmission T [65]. For a pixel (x, y), this can be expressed as:

$$D(x,y) = \frac{1}{T(x,y)} \frac{\sigma_{\text{sam}}(x,y)}{\sigma_{\text{ref}}(x + u_x, y + u_y)}, \tag{17}$$

where u_x and u_y are the displacements of the sample interference pattern in the two orthogonal directions and σ is the standard deviation operator over all diffuser positions for the scanning-based modes or all pixels in the subset window for XST. Although first derived for 2D XSS, the same procedure for the calculation of the dark-field signal can also be applied to the other operational modes such as XSVT [157] and mixed XSVT approaches [135].

Another way to extract the dark-field signal was first proposed for the single-shot XST method [85] and was later extended for the UMPA mode [134]. As outlined in Section 3.1, the reduction of amplitude due to small-angle scattering can be included in a model that expresses the sample speckle pattern in terms of the reference speckle pattern; see Equation (9). A windowed least-squares minimisation procedure delivers here directly the dark-field signal D. The same model is used in the UMPA approach (see Section 3.3), but the signal from several diffuser positions is combined, which allows a higher sensitivity and spatial resolution, also for the dark-field signal [138].

It should be noted that, although the reconstruction approaches in the previous two paragraphs differ, the physical principle of the dark-field contrast generation that they are based on is the same, as pointed out in [163].

An alternative view on the dark-field signal was presented by Wang et al. for the XSS technique [137]. Here, the reconstruction is performed by taking a normalised cross-correlation of the interference patterns in neighbouring pixels. This is done separately for the sample as well as the reference pattern. The normalised maximum correlation coefficient in a pixel (x, y) is defined as the ratio of the maximum sample and the maximum reference correlation coefficients [137]:

$$M(x, y) = M_{\text{sam}}(x, y) / M_{\text{ref}}(x, y). \tag{18}$$

The change in $M(x, y)$ is taken as a measure for the small-angle scattering in the sample, and a reduction of $M(x, y)$ from one pixel to a neighbouring pixel is interpreted as an increased dark-field signal. The absolute dark-field signal in this approach is defined as [137]:

$$D(x, y) = -2 \ln M(x, y). \tag{19}$$

Using 1D scanning, Wang et al. report that this approach can deliver directional dark-field images that include contributions from small-angle scattering as well as the second derivative of the wavefront phase [137]. However, Berujon claims in [163] that the normalised maximum correlation coefficient cannot accurately describe the scattering behaviour of a sample. He furthermore argues that, rather than the second derivative of the phase, the method senses optical phase discontinuities at pixel boundaries that are larger than the pixel size and hence cannot be regarded as a dark-field signal [163].

In another implementation, Wang et al. used an approach based on the reduction of the peak value of the local cross-correlation coefficient between reference and sample signals to calculate the dark-field image for the XST single-shot analysis [89]. The dark-field signal is here defined as $D = 1 - \gamma^{\text{max}}$, where γ^{max} is the maximum (peak) value of the cross-correlation coefficient, which is obtained for each pixel from the zero-normalised cross-correlation of the sample and reference subset windows.

The correlation coefficient was also used in a dark-field approach proposed for 1D XSS dark-field tomography [87,129,136]. The sample speckle pattern is modelled as the convolution of the reference speckle pattern and the optical transfer function of the specimen. The latter can be approximated by taking into account the phase shift as well as the scattering in the sample, where the scattering is modelled as Gaussian and isotropic. Cross-correlation is performed between the sample arrays and the reference arrays. From these considerations and with some further approximations, the maximum of the cross-correlation coefficient can be expressed as [87]:

$$\gamma^{\text{max}} = \exp\left(-\frac{8\pi^4 d^2 \sigma^2}{\zeta_{\text{speckle}}^2}\right), \tag{20}$$

where d is the propagation distance, σ^2 the second moment of the scattering angle distribution and ζ_{speckle} the average speckle size, which can be estimated from the position of the maximum of the speckle pattern power spectrum. This can be rearranged to obtain the dark-field signal D:

$$D = \sigma^2 = \frac{-\zeta_{\text{speckle}}^2}{8\pi^4 d^2} \ln\left(\gamma^{\text{max}}\right) \tag{21}$$

The model of the sample interference pattern as a convolution of the sample scattering distribution and the reference pattern is also used in [157]. However, here the full scattering distribution is analysed using iterative methods (see Section 3.2.5), and the transmission, differential phase and dark-field signals are interpreted as its different moments. In this framework, the two second normalised moments can be seen as the characteristic scattering width in the two orthogonal directions, equivalent to the dark-field signal. Further complementary scattering signals corresponding to various physical phenomena can also be obtained with this approach.

5. Translation to Laboratory Sources and High X-ray Energies

The relatively low requirements on the temporal and spatial coherence of the X-ray beam [133] make speckle-based imaging an ideal candidate for the application at laboratory-based systems with conventional X-ray tubes. The translation of the speckle-based technique to a laboratory source was first demonstrated with the single-shot XST technique [85] at a liquid metal-jet source (Excillum) [164]. Transmission, differential phase and dark-field images were successfully reconstructed for several samples, such as the plastic flower shown in Figure 7. Shortly after, the implementation of the 2D XSS method at the same laboratory source was also reported [86].

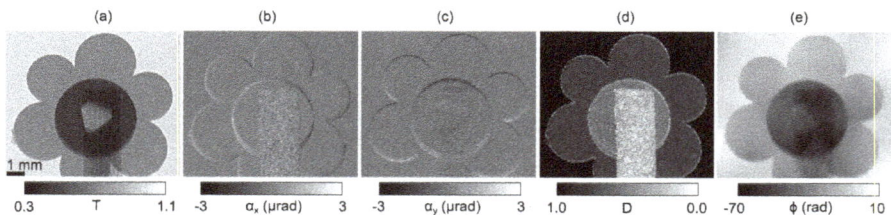

Figure 7. First demonstration of X-ray speckle-based imaging at a laboratory source. (**a**) Transmission, (**b**) differential phase in the horizontal and (**c**) the vertical direction and (**d**) the dark-field signal of a plastic flower on a wooden support could be successfully retrieved. (**e**) Wavefront phase obtained from integration of (**b**,**c**). Reprinted figure with permission from [85]. Copyright (2014) by the American Physical Society.

The liquid metal-jet source used in [85,86] is a laboratory source with relatively high flux and a small spot size that has a polychromatic spectrum dominated by the gallium, indium and tin emission lines of the liquid anode material [164]. However, it has been shown that also conventional micro-focus sources with lower flux and a broader spectrum can be used for speckle-based imaging [87,136]. As the transverse coherence length for conventional sources is lower than for the liquid metal-jet source, it is more challenging to produce X-ray near-field speckle, which relies on scattering and interference effects. It has been demonstrated that a high-visibility reference pattern can also be created by exploiting the absorption of small random structures, e.g., from coarse sandpaper [87] or a "random absorption mask" such as steel wool [136]. The "absorption-speckle" method can be easily applied to a large range of laboratory sources. However, one should be aware that this approach is not speckle imaging in its original definition as the reference pattern is not a speckle pattern based on interference effects. It has been shown that absorption speckle allows the use of high-energy X-rays, for which the contrast of a conventional near-field speckle pattern created by a piece of sandpaper is usually low [136]. On the other hand, recently, near-field speckle-based imaging was demonstrated also using conventional phase speckle with high-energy X-rays from a filtered synchrotron bending magnet beam with a mean energy of 65 keV and 25% bandwidth of the detected spectrum [156]. A speckle pattern of high visibility was achieved in this setup by stacking several sheets of sandpaper.

For a practical implementation of the speckle-based technique with polychromatic X-rays, one should be aware that, as for other imaging methods, artefacts may arise from beam hardening in the specimen, in particular for high-density samples. This has been investigated in detail in a simulation study [133], and the effect has recently been observed experimentally in the dark-field signal of XST measurements conducted at a micro-focus laboratory X-ray source [165].

6. Speckle-Based X-ray Phase-Contrast and Dark-Field Tomography

For many applications, the 2D data alone are not sufficient, and it is essential to obtain quantitative 3D information of the inner density distribution in a specimen. Often, also the 3D scattering distribution is of interest and can give complementary information. As illustrated in the previous sections, speckle-based imaging can provide quantitative phase-contrast signals as well as transmission and dark-field images from a single dataset. The extension from 2D projection imaging to 3D tomography is straightforward. Projections with the sample in the beam are taken at typically a few hundred or thousand different viewing angles of the specimen between $0°$ and $180°$ (or $360°$). Depending on the operational mode (see Section 3), images are acquired at one or several diffuser positions. References without the sample do not need to be taken for each projection, and in principle, it is sufficient to have one reference image at each diffuser position. However, commonly, a few sets of references are recorded to reduce the effects of beam instabilities. For each of the projections, the multimodal image signals are then reconstructed from the acquired raw data. Subsequently, a tomographic reconstruction algorithm, e.g., filtered back-projection [166], is applied to obtain the phase, transmission and dark-field tomograms.

Phase-tomography using the speckle-based technique has been demonstrated both at highly brilliant synchrotron sources as well as in the laboratory. In a first report, the phase and transmission tomograms of a human artery obtained with the single-shot XST mode (see Section 3.1) were shown [89]. The superior sensitivity to density differences of the phase signal over the transmission signal, here between the artery lumen and walls, was observed in a qualitative way; see Figure 10I in Section 7.2.

At around the same time, a quantitative analysis of speckle tomography data was presented from XST measurements at a liquid metal-jet laboratory source [88]. Here, it was shown that the complementary quantitative absorption and refraction information from transmission and phase tomograms, respectively, can be combined for identifying and characterising different materials with similar refraction and absorption properties in a sample; see Figure 10V in Section 7.2.

Furthermore, quantitative phase and dark-field tomographies of a phantom sample were successfully demonstrated using the 1D XSS method (see Section 3.2.2) [129]. However, it should be noted that an object with features oriented mainly along the axis orthogonal to the scanning direction was chosen, and only the refraction signal in the scanning direction was considered in the tomography reconstruction. The sensitivity along the axis opposite the scanning direction is typically lower for the 1D XSS method.

Also, the XSVT approach and the mixed XST-XSVT (see Section 3.3) have been implemented in tomographic mode, and it was shown that complementary absorption, phase and dark-field tomograms of berry samples could successfully be reconstructed [135]. An intelligent interlaced acquisition scheme, similar to the one proposed for grating interferometry in [167], was used here to reduce the number of required diffuser steps for the tomography scan even further by including the information from several subsequent projections. This way, as few as five diffuser positions per projection could be used for the interlaced XSVT tomography when combining the information from additional two projections before and after the projection of interest [135]. With an otherwise identical acquisition scheme, this was reduced to only one diffuser position per projection for the mixed XST-XSVT approach, which in the analysis includes the information from neighbouring pixels in a small window, here 3×3 pixels, centred around the pixel under consideration [135].

A similar interlaced system was applied to the acquisition scheme of sparsely sampled XSS (see Section 3.2.4) [156]. Considering the information from two preceding and two following projections in the analysis of one projection allowed stepping of the diffuser effectively on a 5×5 grid, while only taking five diffuser steps per projection. For projection $p + 5k$ with $k = 0, 1, \ldots, (N-5)/5$; $p = 1, 2, \ldots, 5$, five images were acquired only at the diffuser positions in row p of the 5×5 grid, where N is the total number of projections of the tomography scan. The reference pattern was scanned on a denser grid following the conventional 2D XSS scheme, and the corresponding missing sample frames were obtained by interpolation. High-quality absorption and phase volumes were retrieved with this approach, while the exposure time could be significantly reduced thanks to the sparse sampling and interlaced acquisition scheme.

The translation of UMPA (see Section 3.3) from 2D projection to 3D tomographic mode is currently under way and giving promising results that allow reducing the number of steps down to five or possibly less per projection without the need for interpolation or other computationally expensive preparation of the raw data.

7. Applications of the X-ray Speckle-Based Technique

As X-ray near-field speckle imaging is a versatile, robust and easily implemented technique, it can be expected to find applications in a wide range of fields. Being a relatively young method, a lot of the tremendous potential of speckle-based X-ray imaging has yet to be explored. The main applications of the technique that have been demonstrated so far are illustrated in the following.

7.1. Metrology and Wavefront Sensing

A focus of applications has been the use of X-ray near-field speckle for metrology, optics characterisation and beam phase sensing. The simple and robust experimental arrangement and high angular sensitivity make the speckle-based technique an ideal candidate for metrology. The idea of applying speckle imaging to wavefront measurements and optics characterisation was presented early on in the first publications on the technique [63,65]. In the following years, increasing use of near-field speckle was reported for the characterisation of refractive lenses [93,134,168–171] and X-ray mirrors [90,93,154,155,169,172–175], as well as analysing the local beam wavefront [63,65,169] and measuring the transverse coherence length of the X-ray beam [92,93,176].

For metrology, speckle-based phase-sensing is commonly operated in one of the two modes [65,93]: the differential mode or the self-correlation mode; see Section 3.

Moderately refracting optical elements such as single compound refractive lens (CRL) elements can be analysed using the common differential mode, which is based on acquiring one or more reference interference patterns and one or more sample patterns with the optics in the beam and subsequent reconstruction of the differential phase using one of the available analysis methods (see Section 3). The wavefront phase Φ downstream of the optical element can then be obtained from the differential phase (refraction angle) signal via integration.

The characterisation of a 2D CRL element was first demonstrated using XST [168,169] and 2D XSS [65] in differential mode; see Figure 8I. Furthermore, a 1D parabolic lens made from beryllium was analysed with 1D XSS, as shown in Figure 8II, and the aberrations from the expected wavefront downstream of the lens were retrieved [170]. Just recently, parabolic 1D and 2D CRL elements made from SU-8 polymer material [177] were inspected using the UMPA approach implemented in two configurations, with either a piece of random sandpaper or a periodic phase grating as a phase modulator [134,171]; see Figure 8III. The analysis allowed the sensitive identification and quantification of deviations from the expected refraction behaviour caused by beam damage and shape errors of the lenses, as shown in Figure 8IV.

On the other hand, the self-correlation mode (Section 3.2.3) can be employed to directly measure the absolute effective local wavefront curvature, i.e., the second derivative of the wavefront after passing through the lens [65].

For strongly focussing (or defocussing) optics such as X-ray mirrors, it is essential to use the self-correlation mode (see Section 3.2.3), as in this case the X-ray beam is significantly (de-)magnified by the optical element, and the conventional correlation procedure between sample and reference scans will not succeed in accurately measuring the wavefront distortions. The self-correlation analysis has been applied for the characterisation of mechanically bent and piezo bimorph X-ray mirrors. A single 1D scan of the diffuser allows retrieving the 1D mirror slope [90], and it has been shown that also the 2D slope can be accessed from 1D scanning [154].

(I) Wavefront slope for a 2D beryllium CRL element measured with 2D XSS.

(II) Vertical refraction angle, i.e., wavefront slope (top) and reconstructed phase (bottom) for a 1D beryllium refractive lens measured with 1D XSS. Scale bar corresponds to 0.2 mm.

(III) Horizontal refraction angle for a 2D polymer CRL element measured with UMPA using a random (left) and a periodic (centre) phase modulator to create a reference pattern and comparison with grating interferometry (right).

(IV) Deviation from the expected horizontal (left) and vertical (centre) refraction angle and absolute deviation (right) in the aperture of the 2D polymer CRL in (III), measured with UMPA using a speckle pattern as a wavefront marker.

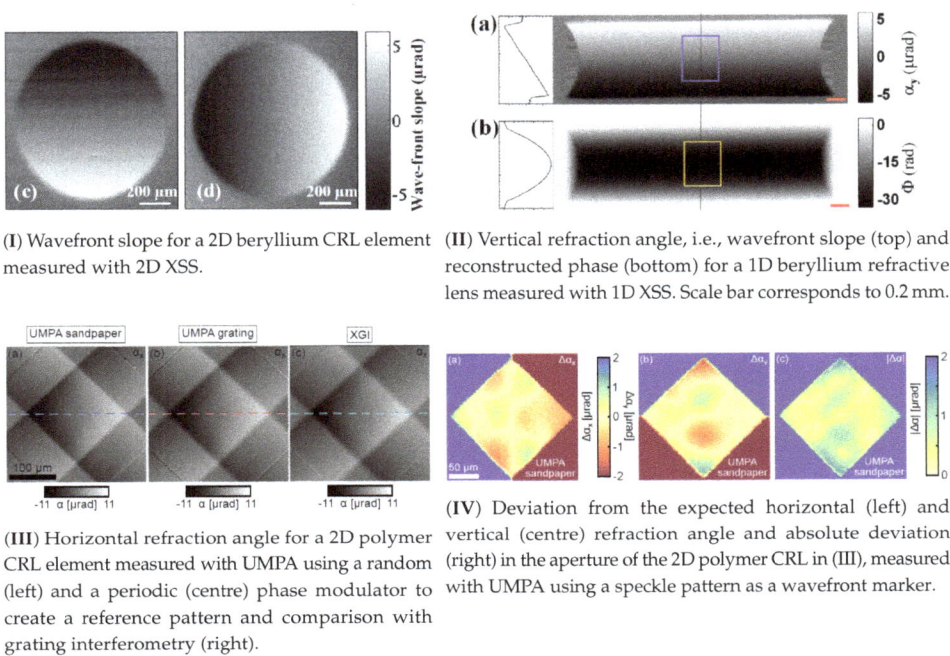

Figure 8. Examples of applications for the characterisation of X-ray refractive lenses. Figures reprinted with permission: (**I**) from [65], Copyright (2012) by the American Physical Society; (**II**) from [170], Copyright (2015) by the Optical Society of America; (**III,IV**) from [171], licensed under CC BY4.0.

As mentioned in Section 3.2.3, measurements can be performed with the diffuser upstream or downstream of the optical element under consideration [155]. The two configurations are illustrated in Figure 9I.

The first configuration gives information about the wavefront distortions caused by the optical element downstream of the diffuser only. This approach is suitable for the characterisation of mirror surfaces in order to detect slope errors. The reflective surface of the mirror and its errors can be analysed directly, as shown in Figure 9II,III. The separation in the plane of the incident wavefront of two rays that are adjacent in the detector plane is obtained from the signal delay given by the correlation procedure. Subsequent integration allows calculating the position of the rays in the diffuser plane, and an iteration process delivers the mirror slope [90]. It was demonstrated that this way slope errors can be accurately determined, allowing precise optimisation of the mirror [90,93,154,155].

The downstream configuration (see Figure 9I) senses the total beam wavefront modulated by all optics in the beam upstream of the diffuser, rather than the properties of an optical element itself. Typically, the local radius of curvature of the wavefront is reconstructed, which is directly related to the second derivative of the wavefront phase Φ (see Equation (14)) [65,90,155,173]. This approach has been applied successfully to the fast, precise optimisation of bimorph mirrors with the aim to obtain a desired beam size and shape [93,155,173–175]; see Figure 9IV. Sensitivities down to 2 nrad have been reported for these applications [155,173]. An in-situ portable metrology device based on this concept has been developed at Diamond Light Source [174]. Its use for the characterisation of elliptical mirrors, the optimisation of bimorph mirrors and mirror alignment has been demonstrated [175].

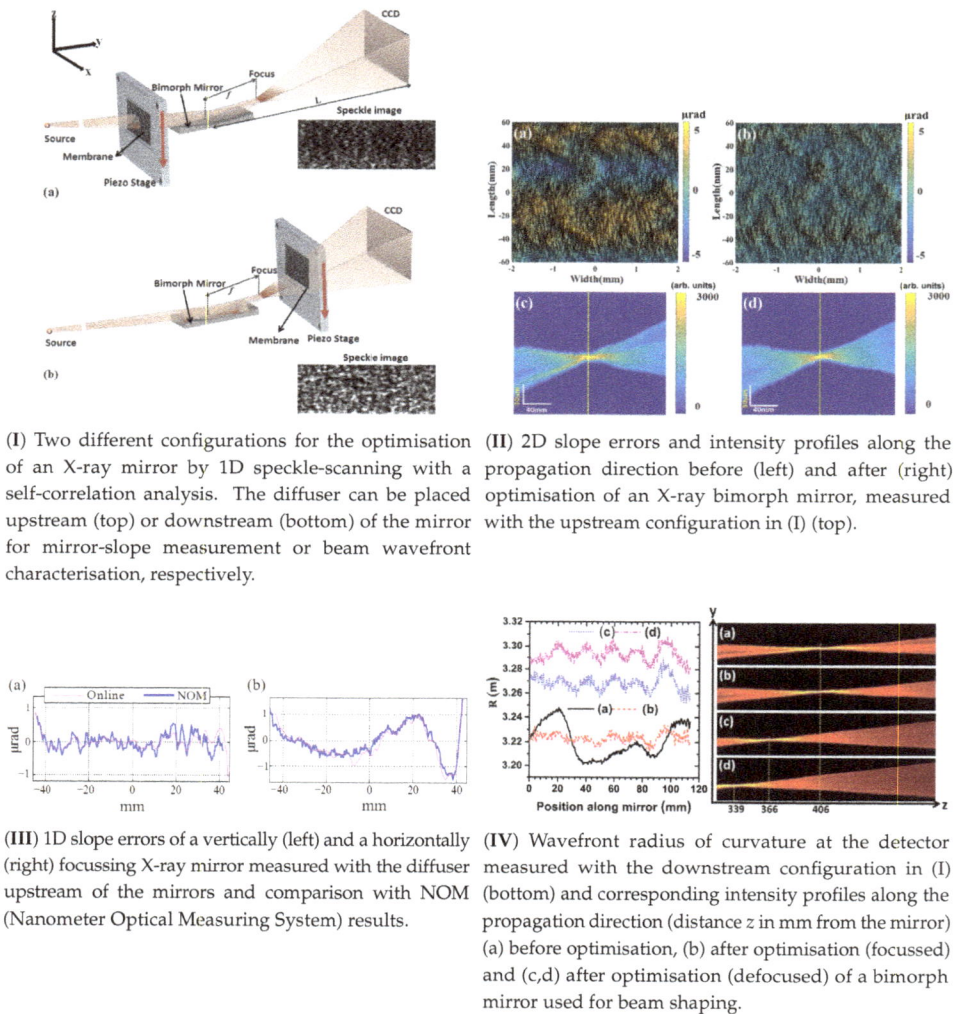

(I) Two different configurations for the optimisation of an X-ray mirror by 1D speckle-scanning with a self-correlation analysis. The diffuser can be placed upstream (top) or downstream (bottom) of the mirror for mirror-slope measurement or beam wavefront characterisation, respectively.

(II) 2D slope errors and intensity profiles along the propagation direction before (left) and after (right) optimisation of an X-ray bimorph mirror, measured with the upstream configuration in (I) (top).

(III) 1D slope errors of a vertically (left) and a horizontally (right) focussing X-ray mirror measured with the diffuser upstream of the mirrors and comparison with NOM (Nanometer Optical Measuring System) results.

(IV) Wavefront radius of curvature at the detector measured with the downstream configuration in (I) (bottom) and corresponding intensity profiles along the propagation direction (distance z in mm from the mirror) (a) before optimisation, (b) after optimisation (focussed) and (c,d) after optimisation (defocused) of a bimorph mirror used for beam shaping.

Figure 9. Examples of applications for X-ray mirror characterisation. Figures reprinted with permission: **(I,II)** from [155], licensed under CC BY 4.0; **(III)** from [90], Copyright (2014) by the Optical Society of America; **(IV)** from [173], Copyright (2015) by the Optical Society of America.

In addition to the investigation of optical elements such as X-ray refractive lenses and mirrors, it can also be of interest to characterise the absolute beam phase without additional beam-shaping optics. This was shown with the XST method in so-called absolute mode [63,169]. In this configuration, images are recorded at two different detector positions along the beam path, and the cross-correlation is performed between these two images without the use of a reference speckle pattern. The recovered local displacement of the speckle pattern relates to the refraction angle, i.e., the first derivative of the wavefront, which can then be integrated to obtain the beam phase. This method, however, requires the beam to remain stable over the course of the two image acquisitions. For cases where the X-ray beam is fluctuating in time, it is more appropriate to record the two images simultaneously and a different variation of this type of measurement was demonstrated for this purpose [91]. The setup consists of a diffuser and one camera with a semi-transparent mirror and scintillator that records an image, but at the same time transmits part of the X-rays, which is recorded by a second camera further downstream. Cross-correlation according to the XST approach between the two simultaneously recorded images delivers the first derivative of the beam phase. Subset distortions can also be taken into account to gather additional information on the second derivative of the beam phase. This approach is particularly suitable for pulsed wavefronts found, e.g., at X-ray free-electron lasers, where the beam profile changes from shot to shot.

Apart from the beam phase, also information about the transverse coherence of the X-ray beam can be obtained using X-ray near-field speckle. This had first been shown in the early days of X-ray near-field speckle by using a colloidal suspension as a diffuser [176] and later with speckle from a filter membrane [92,93]. Only a single exposure of the diffuser is necessary for the analysis, and the transverse coherence length can be retrieved by looking at the Fourier power spectrum of the flat-field corrected speckle interference pattern. As demonstrated in [123], the power spectrum can be decomposed into the 2D scattered intensity distribution and a transfer function. The latter contains contributions from the Talbot effect, the detector response and the partial coherence of the beam. The detector response can be measured, and the Talbot contribution is a known function. When a Gaussian intensity distribution is assumed, the partial coherence term is a function of the transverse coherence lengths in the two orthogonal directions. The known contributions from the detector response and Talbot effect as well as the model of the partial coherence term can be included in a fit function of the angular power spectrum. By fitting of the measured angular power spectrum to the function, using, e.g., a least-squares minimisation procedure, the transverse coherence lengths of the X-ray beam can be determined [92].

7.2. Imaging for Biomedical and Materials Science Applications

Another important and promising area of applications of the speckle-based technique is X-ray phase-contrast and dark-field imaging in particular for biological, biomedical and pre-clinical research as well as materials science.

For biomedical and biological soft-tissue specimens, the phase-contrast signal is of particular interest, as it shows a much higher sensitivity to small density differences than the absorption image for this type of sample. Speckle-based phase-contrast tomography of biomedical and biological specimens has been explored with XST, and the phase tomogram of a human artery was successfully obtained, showing superior contrast compared to the transmission signal [89]; see Figure 10I. Furthermore, 1D XSS was employed to measure a whole fish, and the multimodal images shown in Figure 10III illustrate the complementary character of the different signals, which allows revealing different parts of the sample [87]. The multi-contrast signals of a chicken wing were measured with 1D XSS at a micro-focus laboratory source [87], which is a promising step towards the large-scale and accessible implementation of speckle-based imaging for biomedical applications. The results on the fish and chicken wing [87] as well as the high-contrast scans of different kinds of berries using variations of the XSVT technique [135,156,157] (see Figure 10II) also indicate the potential in the area of food inspection for quality control and foreign body detection. Furthermore, the multimodal UMPA method has proven to be suitable for the investigation of biological samples; see Figure 10IV. It allows flexible tuning of the reconstruction result essential for optimising the trade-off between dose on the specimen and image quality, which is of great importance for biomedical specimens.

(**I**) Volume rendering of the transmission (left) and phase (right) volumes of a human carotid artery obtained with XST tomography.

(**II**) Volume rendering of the dark-field (left) and phase (right) volumes of a juniper berry obtained with interlaced XSVT tomography.

(**III**) Reconstructed absorption, dark-field, differential phase in the vertical and horizontal directions and integrated phase signal (left to right) of a fish obtained with 1D XSS projection imaging (vertical scanning). Scale bar corresponds to 1 mm.

(**IV**) Horizontal, vertical refraction angle, integrated phase, transmission and dark-field (left to right) projections of a small flower bud obtained with UMPA.

(**V**) Material characterisation using XST tomography at a liquid metal-jet laboratory source. Different kinds of plastic in a phantom sample can be distinguished by combining attenuation and refraction information. The diameter of the spheres is 1.5 cm.

(**VI**) Reconstructed transmission (top) and dark-field (bottom) images of a microchip imaged at a laboratory X-ray micro-focus source using 1D XSS with an absorption speckle pattern. Scale bar corresponds to 2 mm.

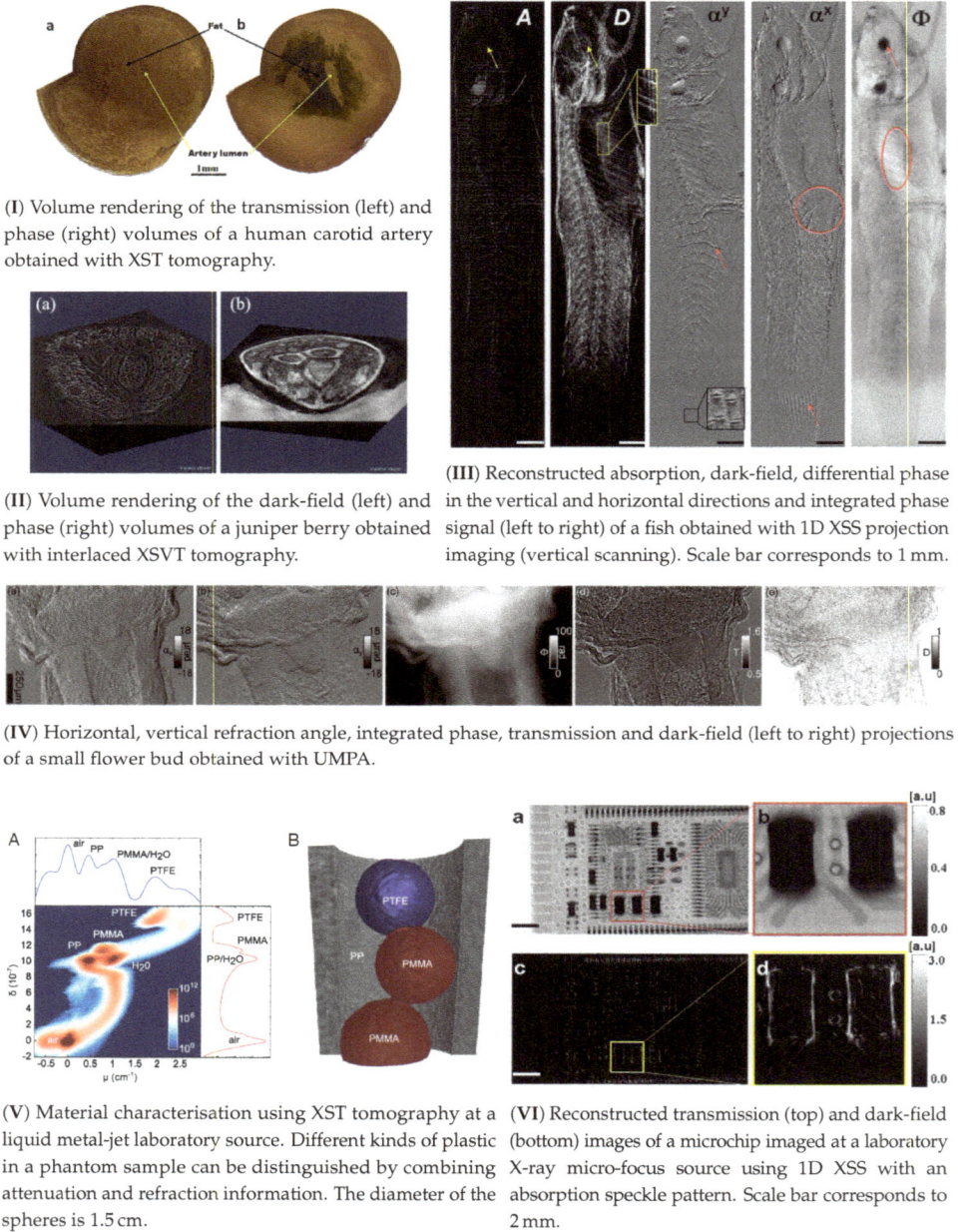

Figure 10. Examples of applications for biomedical and biological imaging and materials science. Figures reprinted with permission: (**I**) from [89], licensed under CC BY 4.0; (**II**) from [135], Copyright (2016) by the American Physical Society; (**III**) from [87], licensed under CC BY 4.0; (**IV**) from [134], licensed under CC BY 4.0; (**V**) from [88]; (**VI**) from [136], licensed under CC BY 4.0.

The complementary character of the various contrast modalities provided by speckle-based imaging can also be exploited for materials science applications. It was demonstrated that XST tomography implemented at laboratory sources can be employed for the identification of different materials in a sample. Several types of plastic were successfully distinguished using the combined information from phase and absorption tomograms [88], as shown in Figure 10V. In another publication, the chip of a computer memory card was imaged using 1D XSS at a laboratory source, and the complementary phase and dark-field signals enabled identifying different components of the chip [136]; see Figure 10VI. These examples suggest the promising potential of the speckle-based technique for the inspection and quality control of electronics as well as identification of different materials in a sample, which can be performed with a simple setup at widely available laboratory sources.

7.3. Other Applications

A different application of X-ray near-field speckle is the use for capturing dynamic processes such as blood flow [126,178–181] and the movement of mouse lungs [182,183]. Here, one makes use of speckle created directly by the specimen under study, and no diffuser is needed. The speckle from blood or from the alveoli of the lung can be used as a marker to track the dynamic process. A windowed cross-correlation analogous to the XST imaging approach is performed between the speckle images recorded at different points in time. From the displacement of the speckle pattern, the speed of the particles can be calculated.

The recent advances in acquisition schemes and reconstruction approaches allow short scan times, flexible tuning of the signal sensitivity and spatial resolution and a straightforward implementation at laboratory sources. Following these developments, it can be expected that the range of applications of the speckle-based technique will increase further in the next few years, and the method will be employed widely at synchrotron and laboratory sources. In particular in the field of biomedical and pre-clinical imaging, speckle-based phase-contrast as well as dark-field imaging have great potential due to their robustness, cost- and dose-effectiveness, amongst others.

8. Conclusions and Outlook

Despite being developed just a few years ago, X-ray near-field speckle-based imaging has already seen rapid development and is receiving rising interest in the X-ray imaging community. The method has been demonstrated in various acquisition and reconstruction modes, catering to different demands on the spatial resolution, angular sensitivity and scan time. The latest advances in the operational modes of the technique offer the opportunity to flexibly tune these properties by adjusting reconstruction and scan parameters.

Developed at synchrotrons, X-ray speckle-based imaging was soon translated to laboratory sources with reduced temporal and spatial coherence without major efforts, making the method available for a wide range of users.

The current applications of the speckle-based technique have been focussed on metrology, optics characterisation and beam phase sensing, for which extremely high sensitivities were achieved. The results obtained with X-ray phase-contrast and dark-field tomography for biomedical applications and materials science indicate the high potential of speckle imaging in these fields. Further applications of speckle-based tomography for multimodal quantitative visualisation of the inner structure of samples are anticipated.

Future work on improving the existing speckle imaging implementations might include the development of alternative diffuser materials that can be adapted to specific experimental setups as well as further optimisation and acceleration of reconstruction algorithms.

The robustness and ease of implementation of X-ray speckle-based imaging has attracted increased interest and extensive research on the technique in the last few years. Following the recent developments, the widespread use of X-ray speckle-based imaging and metrology can be expected for applications in an expanding range of fields.

Acknowledgments: The author gratefully acknowledges Irene Zanette (Diamond Light Source) and Pierre Thibault (University of Southampton) for proofreading the manuscript and providing valuable feedback. The author also highly appreciates their ongoing support and encouragement.

Conflicts of Interest: The author declares no conflict of interest.

Abbreviations

The following abbreviations are used in this manuscript:

1D	one-dimensional
2D	two-dimensional
3D	three-dimensional
FWHM	full width at half maximum
XST	X-ray speckle tracking
XSS	X-ray speckle scanning
XSVT	X-ray speckle-vector tracking
UMPA	unified modulated pattern analysis
CRL	compound refractive lens

References

1. Röntgen, W.C. Über eine neue Art von Strahlen (Vorläufige Mittheilung). *Sitzungsber. Der Würzburger Physik.-Medic. Gesellsch.* **1895**, *9*, 132–141.
2. Röntgen, W.C. Über eine neue Art von Strahlen (Erste Mittheilung). *Ann. Phys. (Berl.)* **1898**, *300*, 1–11.
3. Codman, E.A. Radiograph of fetal arm. *Boston Med. Surg. J.* **1896**, *134*, 327.
4. Editorial. Rare anomalies of the phalanges shown by the Röntgen process. *Boston Med. Surg. J.* **1896**, *134*, 198–199.
5. Editorial. On the application of the Röntgen rays to the diagnosis of arterio-sclerosis. *Boston Med. Surg. J.* **1896**, *134*, 550–551.
6. Spiegel, P.K. The first clinical X-ray made in America—100 years. *AJR Am. J. Roentgenol.* **1995**, *164*, 241–243.
7. Hounsfield, G.N. Computerized transverse axial scanning (tomography): Part 1. Description of system. *Br. J. Radiol.* **1973**, *46*, 1016–1022.
8. Ambrose, J. Computerized transverse axial scanning (tomography): Part 2. Clinical application. *Br. J. Radiol.* **1973**, *46*, 1023–1047.
9. Beckmann, E.C. CT scanning the early days. *Br. J. Radiol.* **2006**, *79*, 5–8.
10. Pollock, H.C. The discovery of synchrotron radiation. *Am. J. Phys.* **1983**, *51*, 278–280.
11. Bilderback, D.H.; Elleaume, P.; Weckert, E. Review of third and next generation synchrotron light sources. *J. Phys. B* **2005**, *38*, S773–S797.
12. Kunz, C. Synchrotron radiation: Third generation sources. *J. Phys. Condens. Matter* **2001**, *13*, 7499–7510.
13. Zernike, F. Phase contrast, a new method for the microscopic observation of transparent objects. *Physica* **1942**, *9*, 686–698.
14. Bonse, U.; Hart, M. An X-ray interferometer. *Appl. Phys. Lett.* **1965**, *6*, 155–156.
15. Fitzgerald, R. Phase-Sensitive X-ray Imaging. *Phys. Today* **2000**, *53*, 23.
16. Lewis, R.A. Medical phase contrast X-ray imaging: Current status and future prospects. *Phys. Med. Biol.* **2004**, *49*, 3573.
17. Momose, A. Recent Advances in X-ray Phase Imaging. *Jpn. J. Appl. Phys.* **2005**, *44*, 6355.
18. Betz, O.; Wegst, U.; Weide, D.; Heethoff, M.; Helfen, L.; Lee, W.-K.; Cloetens, P. Imaging applications of synchrotron X-ray phase-contrast microtomography in biological morphology and biomaterials science. I. General aspects of the technique and its advantages in the analysis of millimetre-sized arthropod structure. *J. Microsc.* **2007**, *227*, 51–71.
19. Williams, I.; Siu, K.; Runxuan, G.; He, X.; Hart, S.; Styles, C.; Lewis, R. Towards the clinical application of X-ray phase contrast imaging. *Eur. J. Radiol.* **2008**, *68*, S73–S77.
20. Zhou, S.-A.; Brahme, A. Development of phase-contrast X-ray imaging techniques and potential medical applications. *Phys. Med.* **2008**, *24*, 129–148.

21. Bravin, A.; Coan, P.; Suortti, P. X-ray phase-contrast imaging: From pre-clinical applications towards clinics. *Phys. Med. Biol.* **2013**, *58*, R1.

22. Coan, P.; Bravin, A.; Tromba, G. Phase-contrast X-ray imaging of the breast: Recent developments towards clinics. *J. Phys. D* **2013**, *46*, 494007.

23. Koehler, T.; Daerr, H.; Martens, G.; Kuhn, N.; Löscher, S.; van Stevendaal, U.; Roessl, E. Slit-scanning differential X-ray phase-contrast mammography: Proof-of-concept experimental studies. *Med. Phys.* **2015**, *42*, 1959–1965.

24. Horn, F.; Hauke, C.; Lachner, S.; Ludwig, V.; Pelzer, G.; Rieger, J.; Schuster, M.; Seifert, M.; Wandner, J.; Wolf, A.; et al. High-energy X-ray grating-based phase-contrast radiography of human anatomy. *Proc. SPIE* **2016**, *9783*.

25. Momose, A.; Yashiro, W.; Kido, K.; Kiyohara, J.; Makifuchi, C.; Ito, T.; Nagatsuka, S.; Honda, C.; Noda, D.; Hattori, T.; et al. X-ray phase imaging: From synchrotron to hospital. *Philos. Trans. Royal Soc. A* **2014**, *372*, 20130023.

26. Cloetens, P.; Pateyron-Salomé, M.; Buffière, J.Y.; Peix, G.; Baruchel, J.; Peyrin, F.; Schlenker, M. Observation of microstructure and damage in materials by phase sensitive radiography and tomography. *J. Appl. Phys.* **1997**, *81*, 5878–5886.

27. Stevenson, A.W.; Gureyev, T.E.; Paganin, D.; Wilkins, S.W.; Weitkamp, T.; Snigirev, A.; Rau, C.; Snigireva, I.; Youn, H.S.; Dolbnya, I.P.; et al. Phase-contrast X-ray imaging with synchrotron radiation for materials science applications. *Nucl. Instr. Meth. Phys. Res. B* **2003**, *199*, 427–435.

28. Zoofan, B.; Kim, J.-Y.; Rokhlin, S.I.; Frankel, G.S. Phase-contrast X-ray imaging for nondestructive evaluation of materials. *J. Appl. Phys.* **2006**, *100*, 014502.

29. Mayo, S.C.; Stevenson, A.W.; Wilkins, S.W. In-Line Phase-Contrast X-ray Imaging and Tomography for Materials Science. *Materials* **2012**, *5*, 937–965.

30. Weitkamp, T.; Nöhammer, B.; Diaz, A.; David, C.; Ziegler, E. X-ray wavefront analysis and optics characterization with a grating interferometer. *Appl. Phys. Lett.* **2005**, *86*, 054101.

31. Engelhardt, M.; Baumann, J.; Schuster, M.; Kottler, C.; Pfeiffer, F.; Bunk, O.; David, C. Inspection of refractive X-ray lenses using high-resolution differential phase contrast imaging with a microfocus X-ray source. *Rev. Sci. Instrum.* **2007**, *78*, 093707.

32. Diaz, A.; Mocuta, C.; Stangl, J.; Keplinger, M.; Weitkamp, T.; Pfeiffer, F.; David, C.; Metzger, T.H.; Bauer, G. Coherence and wavefront characterization of Si-111 monochromators using double-grating interferometry. *J. Synchrotron Radiat.* **2010**, *17*, 299–307,

33. Rutishauser, S.; Zanette, I.; Weitkamp, T.; Donath, T.; David, C. At-wavelength characterization of refractive X-ray lenses using a two-dimensional grating interferometer. *Appl. Phys. Lett.* **2011**, *99*, 221104.

34. Wang, H.; Sawhney, K.; Berujon, S.; Ziegler, E.; Rutishauser, S.; David, C. X-ray wavefront characterization using a rotating shearing interferometer technique. *Opt. Express* **2011**, *19*, 16550–16559.

35. Berujon, S.; Ziegler, E. Grating-based at-wavelength metrology of hard X-ray reflective optics. *Opt. Lett.* **2012**, *37*, 4464–4466.

36. Kayser, Y.; David, C.; Flechsig, U.; Krempasky, J.; Schlott, V.; Abela, R. X-ray grating interferometer for in situ and at-wavelength wavefront metrology. *J. Synchrotron Radiat.* **2017**, *24*, 150–162.

37. Kewish, C.M.; Guizar-Sicairos, M.; Liu, C.; Qian, J.; Shi, B.; Benson, C.; Khounsary, A.M.; Vila-Comamala, J.; Bunk, O.; Fienup, J.R.; et al. Reconstruction of an astigmatic hard X-ray beam and alignment of K-B mirrors from ptychographic coherent diffraction data. *Opt. Express* **2010**, *18*, 23420–23427.

38. Kewish, C.M.; Thibault, P.; Dierolf, M.; Bunk, O.; Menzel, A.; Vila-Comamala, J.; Jefimovs, K.; Pfeiffer, F. Ptychographic characterization of the wavefield in the focus of reflective hard X-ray optics. *Ultramicroscopy* **2010**, *110*, 325–329.

39. Vila-Comamala, J.; Diaz, A.; Guizar-Sicairos, M.; Mantion, A.; Kewish, C.M.; Menzel, A.; Bunk, O.; David, C. Characterization of high-resolution diffractive X-ray optics by ptychographic coherent diffractive imaging. *Opt. Express* **2011**, *19*, 21333–21344.

40. Schropp, A.; Hoppe, R.; Meier, V.; Patommel, J.; Seiboth, F.; Lee, H.J.; Nagler, B.; Galtier, E.C.; Arnold, B.; Zastrau, U.; et al. Full spatial characterization of a nanofocused X-ray free-electron laser beam by ptychographic imaging. *Sci. Rep.* **2013**, *3*, 1633.

41. Momose, A.; Takeda, T.; Itai, Y.; Hirano, K. Phase-contrast X-ray computed tomography for observing biological soft tissues. *Nat. Med.* **1996**, *2*, 473–475.

42. Förster, E.; Goetz, K.; Zaumseil, P. Double crystal diffractometry for the characterization of targets for laser fusion experiments. *Krist. Tech.* **1980**, *15*, 937–945.

43. Davis, T.J.; Gao, D.; Gureyev, T.E.; Stevenson, A.W.; Wilkins, S.W. Phase-contrast imaging of weakly absorbing materials using hard X-rays. *Nature* **1995**, *373*, 595–598.

44. Chapman, D.; Thomlinson, W.; Johnston, R.E.; Washburn, D.; Pisano, E.; Gmür, N.; Zhong, Z.; Menk, R.; Arfelli, F.; Sayers, D. Diffraction enhanced X-ray imaging. *Phys. Med. Biol.* **1997**, *42*, 2015.

45. Pagot, E.; Cloetens, P.; Fiedler, S.; Bravin, A.; Coan, P.; Baruchel, J.; Härtwig, J.; Thomlinson, W. A method to extract quantitative information in analyzer-based X-ray phase contrast imaging. *Appl. Phys. Lett.* **2003**, *82*, 3421–3423.

46. Bravin, A. Exploiting the X-ray refraction contrast with an analyser: The state of the art. *J. Phys. D* **2003**, *36*, A24.

47. Snigirev, A.; Snigireva, I.; Kohn, V.; Kuznetsov, S.; Schelokov, I. On the possibilities of X-ray phase contrast microimaging by coherent high-energy synchrotron radiation. *Rev. Sci. Instrum.* **1995**, *66*, 5486–5492.

48. Wilkins, S.W.; Gureyev, T.E.; Gao, D.; Pogany, A.; Stevenson, A.W. Phase-contrast imaging using polychromatic hard X-rays. *Nature* **1996**, *384*, 335–338.

49. Cloetens, P.; Barrett, R.; Baruchel, J.; Guigay, J.-P.; Schlenker, M. Phase objects in synchrotron radiation hard X-ray imaging. *J. Phys. D* **1996**, *29*, 133–146.

50. Nugent, K.A.; Gureyev, T.E.; Cookson, D.F.; Paganin, D.; Barnea, Z. Quantitative Phase Imaging Using Hard X-rays. *Phys. Rev. Lett.* **1996**, *77*, 2961–2964.

51. Cloetens, P.; Ludwig, W.; Baruchel, J.; Dyck, D.V.; Landuyt, J.V.; Guigay, J.-P.; Schlenker, M. Holotomography: Quantitative phase tomography with micrometer resolution using hard synchrotron radiation X-rays. *Appl. Phys. Lett.* **1999**, *75*, 2912–2914.

52. Paganin, D.; Mayo, S.C.; Gureyev, T.E.; Miller, P.R.; Wilkins, S.W. Simultaneous phase and amplitude extraction from a single defocused image of a homogeneous object. *J. Microsc.* **2002**, *206*, 33–40.

53. David, C.; Nöhammer, B.; Solak, H.H.; Ziegler, E. Differential X-ray phase contrast imaging using a shearing interferometer. *Appl. Phys. Lett.* **2002**, *81*, 3287–3289.

54. Momose, A.; Kawamoto, S.; Koyama, I.; Hamaishi, Y.; Takai, K.; Suzuki, Y. Demonstration of X-ray Talbot Interferometry. *Jpn. J. Appl. Phys.* **2003**, *42*, L866–L868.

55. Weitkamp, T.; Diaz, A.; David, C.; Pfeiffer, F.; Stampanoni, M.; Cloetens, P.; Ziegler, E. X-ray phase imaging with a grating interferometer. *Opt. Express* **2005**, *13*, 6296–6304.

56. Pfeiffer, F.; Weitkamp, T.; Bunk, O.; David, C. Phase retrieval and differential phase-contrast imaging with low-brilliance X-ray sources. *Nat. Phys.* **2006**, *2*, 258–261.

57. Momose, A.; Yashiro, W.; Maikusa, H.; Takeda, Y. High-speed X-ray phase imaging and X-ray phase tomography with Talbot interferometer and white synchrotron radiation. *Opt. Express* **2009**, *17*, 12540–12545.

58. Olivo, A.; Arfelli, F.; Cantatore, G.; Longo, R.; Menk, R.H.; Pani, S.; Prest, M.; Poropat, P.; Rigon, L.; Tromba, G.; et al. An innovative digital imaging set-up allowing a low-dose approach to phase contrast applications in the medical field. *Med. Phys.* **2001**, *28*, 1610–1619.

59. Olivo, A.; Speller, R. A coded-aperture technique allowing X-ray phase contrast imaging with conventional sources. *Appl. Phys. Lett.* **2007**, *91*, 074106.

60. Olivo, A.; Speller, R. Modelling of a novel X-ray phase contrast imaging technique based on coded apertures. *Phys. Med. Biol.* **2007**, *52*, 6555.

61. Olivo, A.; Speller, R. Image formation principles in coded-aperture based X-ray phase contrast imaging. *Phys. Med. Biol.* **2008**, *53*, 6461.

62. Olivo, A.; Ignatyev, K.; Munro, P.R.T.; Speller, R.D. Noninterferometric phase-contrast images obtained with incoherent X-ray sources. *Appl. Opt.* **2011**, *50*, 1765–1769.

63. Bérujon, S.; Ziegler, E.; Cerbino, R.; Peverini, L. Two-Dimensional X-ray Beam Phase Sensing. *Phys. Rev. Lett.* **2012**, *108*, 158102.

64. Morgan, K.S.; Paganin, D.M.; Siu, K.K.W. X-ray phase imaging with a paper analyzer. *Appl. Phys. Lett.* **2012**, *100*, 124102.

65. Berujon, S.; Wang, H.; Sawhney, K. X-ray multimodal imaging using a random-phase object. *Phys. Rev. A* **2012**, *86*, 063813.

66. Pfeiffer, F.; Bech, M.; Bunk, O.; Kraft, P.; Eikenberry, E.F.; Brönnimann, C.; Grünzweig, C.; David, C. Hard-X-ray dark-field imaging using a grating interferometer. *Nat. Mater.* **2008**, *7*, 134–137.

67. Nesterets, Y.I. On the origins of decoherence and extinction contrast in phase-contrast imaging. *Opt. Commun.* **2008**, *281*, 533–542.
68. Yashiro, W.; Terui, Y.; Kawabata, K.; Momose, A. On the origin of visibility contrast in X-ray Talbot interferometry. *Opt. Express* **2010**, *18*, 16890–16901.
69. Baum, T.; Eggl, E.; Malecki, A.; Schaff, F.; Potdevin, G.; Gordijenko, O.; Garcia, E.G.; Burgkart, R.; Rummeny, E.J.; Noël, P.B.; et al. X-ray Dark-Field Vector Radiography—A Novel Technique for Osteoporosis Imaging. *J. Comput. Assist. Tomogr.* **2015**, *39*, 286–289.
70. Ando, M.; Sunaguchi, N.; Shimao, D.; Pan, A.; Yuasa, T.; Mori, K.; Suzuki, Y.; Jin, G.; Kim, J.-K.; Lim, J.-H.; et al. Dark-Field Imaging: Recent developments and potential clinical applications. *Phys. Med.* **2016**, *32*, 1801–1812.
71. Schleede, S.; Meinel, F.G.; Bech, M.; Herzen, J.; Achterhold, K.; Potdevin, G.; Malecki, A.; Adam-Neumair, S.; Thieme, S.F.; Bamberg, F.; et al. Emphysema diagnosis using X-ray dark-field imaging at a laser-driven compact synchrotron light source. *Proc. Natl. Acad. Sci. USA* **2012**, *109*, 17880–17885.
72. Meinel, F.G.; Yaroshenko, A.; Hellbach, K.; Bech, M.; Müller, M.; Velroyen, A.; Bamberg, F.; Eickelberg, O.; Nikolaou, K.; Reiser, M.F.; et al. Improved Diagnosis of Pulmonary Emphysema Using In Vivo Dark-Field Radiography. *Investig. Radiol.* **2014**, *51*, 653–658.
73. Yaroshenko, A.; Hellbach, K.; Yildirim, A.Ö.; Conlon, T.M.; Fernandez, I.E.; Bech, M.; Velroyen, A.; Meinel, F.G.; Auweter, S.; Reiser, M.; et al. Improved In vivo Assessment of Pulmonary Fibrosis in Mice using X-ray Dark-Field Radiography. *Sci. Rep.* **2015**, *5*, 17492.
74. Yaroshenko, A.; Pritzke, T.; Koschlig, M.; Kamgari, N.; Willer, K.; Gromann, L.; Auweter, S.; Hellbach, K.; Reiser, M.; Eickelberg, O.; et al. Visualization of neonatal lung injury associated with mechanical ventilation using X-ray dark-field radiography. *Sci. Rep.* **2016**, *6*, 24269.
75. Hellbach, K.; Yaroshenko, A.; Willer, K.; Pritzke, T.; Baumann, A.; Hesse, N.; Auweter, S.; Reiser, M.F.; Eickelberg, O.; Pfeiffer, F.; et al. Facilitated Diagnosis of Pneumothoraces in Newborn Mice Using X-ray Dark-Field Radiography. *Invest. Radiol.* **2016**, *49*, 597–601.
76. Noël, P.B.; Willer, K.; Fingerle, A.A.; Gromann, L.B.; Marco, F.D.; Scherer, K.H.; Herzen, J.; Achterhold, K.; Gleich, B.; Münzel, D.; et al. First experience with X-ray dark-field radiography for human chest imaging (Conference Presentation). *Proc. SPIE* **2017**, *10132*, 10132.
77. Gromann, L.B.; De Marco, F.; Willer, K.; Noël, P.B.; Scherer, K.; Renger, B.; Gleich, B.; Achterhold, K.; Fingerle, A.A.; Münzel, D.; et al. In-vivo X-ray Dark-Field Chest Radiography of a Pig. *Sci. Rep.* **2017**, *7*, 4807.
78. Revol, V.; Jerjen, I.; Kottler, C.; Schütz, P.; Kaufmann, R.; Lüthi, T.; Sennhauser, U.; Straumann, U.; Urban, C. Sub-pixel porosity revealed by X-ray scatter dark field imaging. *J. Appl. Phys.* **2011**, *110*, 044912.
79. Revol, V.; Plank, B.; Kaufmann, R.; Kastner, J.; Kottler, C.; Neels, A. Laminate fibre structure characterisation of carbon fibre-reinforced polymers by X-ray scatter dark field imaging with a grating interferometer. *NDT E Int.* **2013**, *58*, 64–71.
80. Lauridsen, T.; Willner, M.; Bech, M.; Pfeiffer, F.; Feidenhans'l, R. Detection of sub-pixel fractures in X-ray dark-field tomography. *Appl. Phys. A* **2015**, *121*, 1243–1250.
81. Yang, F.; Prade, F.; Griffa, M.; Jerjen, I.; Bella, C.D.; Herzen, J.; Sarapata, A.; Pfeiffer, F.; Lura, P. Dark-field X-ray imaging of unsaturated water transport in porous materials. *Appl. Phys. Lett.* **2014**, *105*, 154105.
82. Prade, F.; Chabior, M.; Malm, F.; Grosse, C.U.; Pfeiffer, F. Observing the setting and hardening of cementitious materials by X-ray dark-field radiography. *Cem. Concr. Res.* **2015**, *74*, 19–25.
83. Prade, F.; Fischer, K.; Heinz, D.; Meyer, P.; Mohr, J.; Pfeiffer, F. Time resolved X-ray Dark-Field Tomography Revealing Water Transport in a Fresh Cement Sample. *Sci. Rep.* **2016**, *6*, 29108.
84. Schaff, F.; Bachmann, A.; Zens, A.; Zäh, M.F.; Pfeiffer, F.; Herzen, J. Grating-based X-ray dark-field computed tomography for the characterization of friction stir welds: A feasibility study. *Mater. Charact.* **2017**, *129*, 143–148.
85. Zanette, I.; Zhou, T.; Burvall, A.; Lundström, U.; Larsson, D.H.; Zdora, M.-C.; Thibault, P.; Pfeiffer, F.; Hertz, H.M. Speckle-Based X-ray Phase-Contrast and Dark-Field Imaging with a Laboratory Source. *Phys. Rev. Lett.* **2014**, *112*, 253903.
86. Zhou, T.; Zanette, I.; Zdora, M.-C.; Lundström, U.; Larsson, D.H.; Hertz, H.M.; Pfeiffer, F.; Burvall, A. Speckle-based X-ray phase-contrast imaging with a laboratory source and the scanning technique. *Opt. Lett.* **2015**, *40*, 2822–2825.

87. Wang, H.; Kashyap, Y.; Sawhney, K. From synchrotron radiation to lab source: Advanced speckle-based X-ray imaging using abrasive paper. *Sci. Rep.* **2016**, *6*, 20476.
88. Zanette, I.; Zdora, M.-C.; Zhou, T.; Burvall, A.; Larsson, D.H.; Thibault, P.; Hertz, H.M.; Pfeiffer, F. X-ray microtomography using correlation of near-field speckles for material characterization. *Proc. Natl. Acad. Sci. USA* **2015**, *112*, 12569–12573.
89. Wang, H.; Berujon, S.; Herzen, J.; Atwood, R.; Laundy, D.; Hipp, A.; Sawhney, K. X-ray phase contrast tomography by tracking near-field speckle. *Sci. Rep.* **2015**, *5*, 8762.
90. Berujon, S.; Wang, H.; Alcock, S.; Sawhney, K. At-wavelength metrology of hard X-ray mirror using near-field speckle. *Opt. Express* **2014**, *22*, 6438–6446.
91. Berujon, S.; Ziegler, E.; Cloetens, P. X-ray pulse wavefront metrology using speckle tracking. *J. Synchrotron Radiat.* **2015**, *22*, 886–894.
92. Kashyap, Y.; Wang, H.; Sawhney, K. Two-dimensional transverse coherence measurement of hard-X-ray beams using near-field speckle. *Phys. Rev. A* **2015**, *92*, 033842.
93. Wang, H.; Zhou, T.; Kashyap, Y.; Sawhney, K. Speckle-based at-wavelength metrology of X-ray optics at Diamond Light Source. *Proc. SPIE* **2017**, *10388*, 103880I.
94. Dainty, J.C. (Ed.) *Laser Speckle and Related Phenomena*; Topics in Applied Physics; Springer: Englewood, CO, USA, 1975; Volume 9.
95. Goodman, J.W. *Speckle Phenomena in Optics. Theory and Application*; Roberts and Company: Greenwood Village, CO, USA, 2007.
96. Chellappan, K.V.; Erden, E.; Urey, H. Laser-based displays: A review. *Appl. Opt.* **2010**, *49*, F79–F98,
97. Kuratomi, Y.; Sekiya, K.; Satoh, H.; Tomiyama, T.; Kawakami, T.; Katagiri, B.; Suzuki, Y.; Uchida, T. Speckle reduction mechanism in laser rear projection displays using a small moving diffuser. *J. Opt. Soc. Am. A* **2010**, *27*, 1812–1817.
98. Pan, J.-W.; Shih, C.-H. Speckle reduction and maintaining contrast in a LASER pico-projector using a vibrating symmetric diffuser. *Opt. Express* **2014**, *22*, 6464–6477.
99. Liba, O.; Lew, M.D.; SoRelle, E.D.; Dutta, R.; Sen, D.; Moshfeghi, D.M.; Chu, S.; de la Zerda, A. Speckle-modulating optical coherence tomography in living mice and humans. *Nat. Commun.* **2017**, *8*, 15845.
100. Schmitt, J.M.; Xiang, S.H.; Yung, K.M. Speckle in optical coherence tomography. *J. Biomed. Opt.* **1999**, *4*, 95–105.
101. Bates, R.H.T. Astronomical speckle imaging. *Phys. Rep.* **1982**, *90*, 203–297.
102. Horch, E. Speckle imaging in astronomy. *Int. J. Imaging Syst. Technol.* **1995**, *6*, 401–417.
103. Dainty, J.C. Stellar Speckle Interferometry. In *Laser Speckle and Related Phenomena*; Topics in Applied Physics; Dainty, J.C., Ed.; Springer: Englewood, CO, USA, 1975; Volume 9, Chapter 7.
104. Jones, R.; Wykes, C. *Holographic and Speckle Interferometry*; Cambridge Studies in Modern Optics; Cambridge University Press: Cambridge, UK, 1989.
105. Høgmoen, K.; Pedersen, H.M. Measurement of small vibrations using electronic speckle pattern interferometry: Theory. *J. Opt. Soc. Am.* **1977**, *67*, 1578–1583.
106. Ennos, A.E. Speckle interferometry. In *Laser Speckle and Related Phenomena*; Topics in Applied Physics; Dainty, J.C., Ed.; Springer: Englewood, CO, USA, 1975; Volume 9, Chapter 6.
107. Sharp, B. Electronic speckle pattern interferometry (ESPI). *Opt. Laser Eng.* **1989**, *11*, 241–255.
108. Yang, L.; Xie, X.; Zhu, L.; Wu, S.; Wang, Y. Review of electronic speckle pattern interferometry (ESPI) for three dimensional displacement measurement. *Chin. J. Mech. Eng. En.* **2014**, *27*, 1–13.
109. Aizu, Y.; Asakura, T. Bio-speckles. In *Trends in Optics*; Consortini, A., Ed.; Lasers and Optical Engineering, Academic Press: San Diego, CA, USA, 1996; Chapter 2, pp. 27–49.
110. Aizu, Y.; Asakura, T. Bio-speckle phenomena and their application to the evaluation of blood flow. *Opt. Laser Technol.* **1991**, *23*, 205–219.
111. Fujii, H.; Asakura, T.; Nohira, K.; Shintomi, Y.; Ohura, T. Blood flow observed by time-varying laser speckle. *Opt. Lett.* **1985**, *10*, 104–106.
112. Zheng, B.; Pleass, C.M.; Ih, C.S. Feature information extraction from dynamic biospeckle. *Appl. Opt.* **1994**, *33*, 231–237.
113. Rabal, H.; Braga, R. (Eds.) *Dynamic Laser Speckle and Applications*; Optical Science and Engineering, CRC Press: Boca Raton, FL, USA, 2008.

114. Boas, D.A.; Dunn, A.K. Laser speckle contrast imaging in biomedical optics. *J. Biomed. Opt.* **2010**, *15*, 011109.

115. Mohon, N.; Rodemann, A. Laser Speckle for Determining Ametropia and Accommodation Response of the Eye. *Appl. Opt.* **1973**, *12*, 783–787.

116. Giglio, M.; Carpineti, M.; Vailati, A. Space Intensity Correlations in the Near Field of the Scattered Light: A Direct Measurement of the Density Correlation Function $g(r)$. *Phys. Rev. Lett.* **2000**, *85*, 1416–1419.

117. Giglio, M.; Carpineti, M.; Vailati, A.; Brogioli, D. Near-field intensity correlations of scattered light. *Appl. Opt.* **2001**, *40*, 4036–4040.

118. Brogioli, D.; Vailati, A.; Giglio, M. Heterodyne near-field scattering. *Appl. Phys. Lett.* **2002**, *81*, 4109–4111.

119. Giglio, M.; Brogiol, D.; Potenza, M.A.C.; Vailati, A. Near field scattering. *Phys. Chem. Chem. Phys.* **2004**, *6*, 1547–1550.

120. Cerbino, R. Correlations of light in the deep Fresnel region: An extended Van Cittert and Zernike theorem. *Phys. Rev. A* **2007**, *75*, 053815.

121. Gatti, A.; Magatti, D.; Ferri, F. Three-dimensional coherence of light speckles: Theory. *Phys. Rev. A* **2008**, *78*, 063806.

122. Magatti, D.; Gatti, A.; Ferri, F. Three-dimensional coherence of light speckles: Experiment. *Phys. Rev. A* **2009**, *79*, 053831.

123. Cerbino, R.; Peverini, L.; Potenza, M.A.C.; Robert, A.; Bösecke, P.; Giglio, M. X-ray-scattering information obtained from near-field speckle. *Nat. Phys.* **2008**, *4*, 238–243.

124. Suzuki, Y.; Yagi, N.; Uesugi, K. X-ray refraction-enhanced imaging and a method for phase retrieval for a simple object. *J. Synchrotron Radiat.* **2002**, *9*, 160–165.

125. Kitchen, M.J.; Paganin, D.; Lewis, R.A.; Yagi, N.; Uesugi, K.; Mudie, S.T. On the origin of speckle in X-ray phase contrast images of lung tissue. *Phys. Med. Biol.* **2004**, *49*, 4335.

126. Kim, G.B.; Lee, S.J. X-ray PIV measurements of blood flows without tracer particles. *Exp. Fluids* **2006**, *41*, 195–200,

127. Federation of European Producers of Abrasives. FEPA P-Grit Sizes Coated Abrasives. Available online: http://www.fepa-abrasives.org/Abrasiveproducts/Grains/Pgritsizescoated.aspx (accessed on 16 February 2018).

128. Aloisio, I.A.; Paganin, D.M.; Wright, C.A.; Morgan, K.S. Exploring experimental parameter choice for rapid speckle-tracking phase-contrast X-ray imaging with a paper analyzer. *J. Synchrotron Radiat.* **2015**, *22*, 1279–1288.

129. Wang, H.; Kashyap, Y.; Sawhney, K. Quantitative X-ray dark-field and phase tomography using single directional speckle scanning technique. *Appl. Phys. Lett.* **2016**, *108*.

130. Goodman, J.W., Statistical properties of laser speckle patterns. In *Laser Speckle and Related Phenomena*; Topics in Applied Physics; Dainty, J.C., Ed.; Springer: Englewood, CO, USA, 1984; Volume 9, Chapter 2.

131. Alexander, T.L.; Harvey, J.E.; Weeks, A.R. Average speckle size as a function of intensity threshold level: Comparison of experimental measurements with theory. *Appl. Opt.* **1994**, *33*, 8240–8250.

132. Hamed, A. Recognition of direction of new apertures from the elongated speckle images: Simulation. *Opt. Photonics J.* **2013**, *3*, 250–258.

133. Zdora, M.-C.; Thibault, P.; Pfeiffer, F.; Zanette, I. Simulations of X-ray speckle-based dark-field and phase-contrast imaging with a polychromatic beam. *J. Appl. Phys.* **2015**, *118*, 113105.

134. Zdora, M.-C.; Thibault, P.; Zhou, T.; Koch, F.J.; Romell, J.; Sala, S.; Last, A.; Rau, C.; Zanette, I. X-ray Phase-Contrast Imaging and Metrology through Unified Modulated Pattern Analysis. *Phys. Rev. Lett.* **2017**, *118*, 203903.

135. Berujon, S.; Ziegler, E. X-ray Multimodal Tomography Using Speckle-Vector Tracking. *Phys. Rev. Appl.* **2016**, *5*, 044014.

136. Wang, H.; Kashyap, Y.; Cai, B.; Sawhney, K. High energy X-ray phase and dark-field imaging using a random absorption mask. *Sci. Rep.* **2016**, *6*, 30581.

137. Wang, H.; Kashyap, Y.; Sawhney, K. Hard-X-ray Directional Dark-Field Imaging Using the Speckle Scanning Technique. *Phys. Rev. Lett.* **2015**, *114*, 103901.

138. Zdora, M.-C.; Thibault, P.; Rau, C.; Zanette, I. Characterisation of speckle-based X-ray phase-contrast imaging. *J. Phys. Conf. Ser.* **2017**, *849*, 012024.

139. Berujon, S.; Wang, H.; Pape, I.; Sawhney, K. X-ray phase microscopy using the speckle tracking technique. *Appl. Phys. Lett.* **2013**, *102*, 154105.

140. Kottler, C.; David, C.; Pfeiffer, F.; Bunk, O. A two-directional approach for grating based differential phase contrast imaging using hard X-rays. *Opt. Express* **2007**, *15*, 1175–1181.

141. Frankot, R.T.; Chellappa, R. A method for enforcing integrability in shape from shading algorithms. *IEEE Trans. Pattern Anal. Mach. Intell.* **1988**, *10*, 439–451.

142. Harker, M.; O'Leary, P. Least squares surface reconstruction from measured gradient fields. In Proceedings of the IEEE Conference on Computer Vision and Pattern Recognition, Anchorage, AK, USA, 23–28 June 2008; IEEE: Piscataway, New Jersey, USA, 2008; pp. 1–7.

143. Jiang, M.; Wyatt, C.L.; Wang, G. X-ray Phase-Contrast Imaging with Three 2D Gratings. *Int. J. Biomed. Imaging* **2008**, *2008*.

144. Zanette, I.; Weitkamp, T.; Donath, T.; Rutishauser, S.; David, C. Two-Dimensional X-ray Grating Interferometer. *Phys. Rev. Lett.* **2010**, *105*, 248102.

145. Kashyap, Y.; Wang, H.; Sawhney, K. Experimental comparison between speckle and grating-based imaging technique using synchrotron radiation X-rays. *Opt. Express* **2016**, *24*, 18664–18673.

146. Bennett, E.E.; Kopace, R.; Stein, A.F.; Wen, H. A grating-based single-shot X-ray phase contrast and diffraction method for in vivo imaging. *Med. Phys.* **2010**, *37*, 6047–6054.

147. Wen, H.H.; Bennett, E.E.; Kopace, R.; Stein, A.F.; Pai, V. Single-shot X-ray differential phase-contrast and diffraction imaging using two-dimensional transmission gratings. *Opt. Lett.* **2010**, *35*, 1932–1934.

148. Morgan, K.S.; Paganin, D.M.; Siu, K.K.W. Quantitative X-ray phase-contrast imaging using a single grating of comparable pitch to sample feature size. *Opt. Lett.* **2011**, *36*, 55–57.

149. Morgan, K.S.; Paganin, D.M.; Siu, K.K.W. Quantitative single-exposure X-ray phase contrast imaging using a single attenuation grid. *Opt. Express* **2011**, *19*, 19781–19789.

150. Morgan, K.S.; Modregger, P.; Irvine, S.C.; Rutishauser, S.; Guzenko, V.A.; Stampanoni, M.; David, C. A sensitive X-ray phase contrast technique for rapid imaging using a single phase grid analyzer. *Opt. Lett.* **2013**, *38*, 4605–4608.

151. Wang, F.; Wang, Y.; Wei, G.; Du, G.; Xue, Y.; Hu, T.; Li, K.; Deng, B.; Xie, H.; Xiao, T. Speckle-tracking X-ray phase-contrast imaging for samples with obvious edge-enhancement effect. *Appl. Phys. Lett.* **2017**, *111*, 174101.

152. Cloetens, P.; Guigay, J.P.; Martino, C.D.; Baruchel, J.; Schlenker, M. Fractional Talbot imaging of phase gratings with hard x rays. *Opt. Lett.* **1997**, *22*, 1059–1061.

153. Pan, B.; Qian, K.; Xie, H.; Asundi, A. Two-dimensional digital image correlation for in-plane displacement and strain measurement: A review. *Meas. Sci. Technol.* **2009**, *20*, 062001.

154. Wang, H.; Kashyap, Y.; Laundy, D.; Sawhney, K. Two-dimensional *in situ* metrology of X-ray mirrors using the speckle scanning technique. *J. Synchrotron Radiat.* **2015**, *22*, 925–929.

155. Kashyap, Y.; Wang, H.; Sawhney, K. Speckle-based at-wavelength metrology of X-ray mirrors with super accuracy. *Rev. Sci. Instrum.* **2016**, *87*, 052001.

156. Berujon, S.; Ziegler, E. Near-field speckle-scanning-based X-ray tomography. *Phys. Rev. A* **2017**, *95*, 063822.

157. Berujon, S.; Ziegler, E. Near-field speckle-scanning-based X-ray imaging. *Phys. Rev. A* **2015**, *92*, 013837.

158. Modregger, P.; Scattarella, F.; Pinzer, B.R.; David, C.; Bellotti, R.; Stampanoni, M. Imaging the Ultrasmall-Angle X-ray Scattering Distribution with Grating Interferometry. *Phys. Rev. Lett.* **2012**, *108*, 048101.

159. Modregger, P.; Rutishauser, S.; Meiser, J.; David, C.; Stampanoni, M. Two-dimensional ultra-small angle X-ray scattering with grating interferometry. *Appl. Phys. Lett.* **2014**, *105*, 024102.

160. Lucy, L.B. An iterative technique for the rectification of observed distributions. *Astron. J.* **1974**, *79*, 745.

161. Richardson, W.H. Bayesian-Based Iterative Method of Image Restoration*. *J. Opt. Soc. Am.* **1972**, *62*, 55–59.

162. Zhou, T.; Zdora, M.-C.; Zanette, I.; Romell, J.; Hertz, H.M.; Burvall, A. Noise analysis of speckle-based X-ray phase-contrast imaging. *Opt. Lett.* **2016**, *41*, 5490–5493.

163. Berujon, S. Comment on "Hard-X-ray Directional Dark-Field Imaging Using the Speckle Scanning Technique" 2015. Available online: https://doi.org/10.13140/rg.2.1.3975.9608 (accessed on 20 February 2018).

164. Hemberg, O.; Otendal, M.; Hertz, H.M. Liquid-metal-jet anode electron-impact X-ray source. *Appl. Phys. Lett.* **2003**, *83*, 1483–1485.

165. Vittoria, F.A.; Endrizzi, M.; Olivo, A. Retrieving the Ultrasmall-Angle X-ray Scattering Signal with Polychromatic Radiation in Speckle-Tracking and Beam-Tracking Phase-Contrast Imaging. *Phys. Rev. Appl.* **2017**, *7*, 034024.

166. Kak, A.C.; Slaney, M. *Principles of Computerized Tomographic Imaging*; IEEE Press: New York, USA, 1988.

167. Zanette, I.; Bech, M.; Pfeiffer, F.; Weitkamp, T. Interlaced phase stepping in phase-contrast X-ray tomography. *Appl. Phys. Lett.* **2011**, *98*, 094101.

168. Berujon, S.; Wang, H.; Sawhney, K.J.S. At-wavelength metrology using the X-ray speckle tracking technique: Case study of a X-ray compound refractive lens. *J. Phys. Conf. Ser.* **2013**, *425*, 052020.

169. Wang, H.; Berujon, S.; Sutter, J.; Alcock, S.G.; Sawhney, K. At-wavelength metrology of X-ray optics at Diamond Light Source. *Proc. SPIE* **2014**, *9206*, 9206.

170. Wang, H.; Kashyap, Y.; Sawhney, K. Speckle based X-ray wavefront sensing with nanoradian angular sensitivity. *Opt. Express* **2015**, *23*, 23310–23317.

171. Zdora, M.-C.; Zanette, I.; Zhou, T.; Koch, F.J.; Romell, J.; Sala, S.; Last, A.; Ohishi, Y.; Hirao, N.; Rau, C.; et al. At-wavelength optics characterisation via X-ray speckle- and grating-based unified modulated pattern analysis. *Opt. Express* **2018**, *26*, 4989–5004.

172. Sawhney, K.; Alcock, S.; Sutter, J.; Berujon, S.; Wang, H.; Signorato, R. Characterisation of a novel super-polished bimorph mirror. *J. Phys. Conf. Ser.* **2013**, *425*, 052026.

173. Wang, H.; Sutter, J.; Sawhney, K. Advanced in situ metrology for X-ray beam shaping with super precision. *Opt. Express* **2015**, *23*, 1605–1614.

174. Kashyap, Y.; Wang, H.; Sawhney, K. Development of a speckle-based portable device for in situ metrology of synchrotron X-ray mirrors. *J. Synchrotron Radiat.* **2016**, *23*, 1131–1136.

175. Wang, H.; Kashyap, Y.; Zhou, T.; Sawhney, K. Speckle-based portable device for in-situ metrology of X-ray mirrors at Diamond Light Source. *Proc. SPIE* **2017**, *10385*, 1038504.

176. Alaimo, M.D.; Potenza, M.A.C.; Manfredda, M.; Geloni, G.; Sztucki, M.; Narayanan, T.; Giglio, M. Probing the Transverse Coherence of an Undulator X-ray Beam Using Brownian Particles. *Phys. Rev. Lett.* **2009**, *103*, 194805.

177. Nazmov, V.; Reznikova, E.; Mohr, J.; Snigirev, A.; Snigireva, I.; Achenbach, S.; Saile, V. Fabrication and preliminary testing of X-ray lenses in thick SU-8 resist layers. *Microsys. Technol.* **2004**, *10*, 716–721.

178. Kim, G.B.; Lee, S.J. Contrast enhancement of speckle patterns from blood in synchrotron X-ray imaging. *J. Biomech.* **2009**, *42*, 449–454.

179. Irvine, S.C.; Paganin, D.M.; Dubsky, S.; Lewis, R.A.; Fouras, A. Phase retrieval for improved three-dimensional velocimetry of dynamic X-ray blood speckle. *Appl. Phys. Lett.* **2008**, *93*, 153901.

180. Park, H.; Yeom, E.; Lee, S.J. X-ray PIV measurement of blood flow in deep vessels of a rat: An in vivo feasibility study. *Sci. Rep.* **2016**, *6*, 19194.

181. Izadifar, M.; Kelly, M.E.; Peeling, L. Synchrotron speckle-based X-ray phase-contrast imaging for mapping intra-aneurysmal blood flow without contrast agent. *Biomed. Phys. Eng. Express* **2018**, *4*, 015011.

182. Murrie, R.P.; Morgan, K.S.; Maksimenko, A.; Fouras, A.; Paganin, D.M.; Hall, C.; Siu, K.K.W.; Parsons, D.W.; Donnelley, M. Live small-animal X-ray lung velocimetry and lung micro-tomography at the Australian Synchrotron Imaging and Medical Beamline. *J. Synchrotron Radiat.* **2015**, *22*, 1049–1055.

183. Murrie, R.P.; Paganin, D.M.; Fouras, A.; Morgan, K.S. Phase contrast X-ray velocimetry of small animal lungs: Optimising imaging rates. *Biomed. Opt. Express* **2016**, *7*, 79–92.

Journal of *Imaging*

MDPI

Article

Improved Reconstruction Technique for Moiré Imaging Using an X-Ray Phase-Contrast Talbot–Lau Interferometer

Maria Seifert *, Michael Gallersdörfer, Veronika Ludwig, Max Schuster, Florian Horn, Georg Pelzer, Jens Rieger, Thilo Michel and Gisela Anton

ECAP-Novel Detectors/Medical Physics, Friedrich-Alexander-University Erlangen-Nuremberg, Erwin-Rommel-Strasse 1, 91058 Erlangen, Germany; michael.gallersdoerfer@fau.de (M.G.); veronika.ludwig@fau.de (V.L.); max.schuster@fau.de (M.S.); florian.horn@fau.de (F.H.); georg.pelzer@fau.de (G.P.); jens.rieger@fau.de (J.R.); thilo.michel@fau.de (T.M.); gisela.anton@fau.de (G.A.)
* Correspondence: maria.seifert@fau.de

Received: 21 February 2018; Accepted: 26 April 2018; Published: 1 May 2018

Abstract: X-ray phase-contrast imaging is a promising method for medical imaging and non-destructive testing. Information about the attenuation, small-angle scattering and phase-shifting properties of an object can be gained simultaneously in three image modalities using a Talbot–Lau interferometer. This is a highly sensitive approach for retrieving this information. Nevertheless, until now, Talbot–Lau interferometry has been a time-consuming process due to image acquisition by phase-stepping procedures. Thus, methods to accelerate the image acquisition process in Talbot–Lau interferometry would be desirable. This is especially important for medical applications to avoid motion artifacts. In this work, the Talbot–Lau interferometry is combined with the moiré imaging approach. Firstly, the reconstruction algorithm of moiré imaging is improved compared to the standard reconstruction methods in moiré imaging that have been published until now. Thus, blurring artifacts resulting from the reconstruction in the frequency domain can be reduced. Secondly, the improved reconstruction algorithm allows for reducing artifacts in the reconstructed images resulting from inhomogeneities of the moiré pattern in large fields of view. Hence, the feasibility of differential phase-contrast imaging with regard to the integration into workflows in medical imaging and non-destructive testing is improved considerably. New fields of applications can be gained due to the accelerated imaging process—for example, live imaging in medical applications.

Keywords: X-ray imaging; phase-contrast; Talbot–Lau; image reconstruction; spectrum analysis; fringe analysis; moiré imaging

1. Introduction

In conventional X-ray imaging, the attenuation image of an object is acquired. Thus, for example, in medical imaging, a good bone-tissue contrast can be achieved. Nevertheless, in recent years, many approaches have been made to get further information about the object by analyzing the behavior of the wave passing the object. These techniques can be divided into five main categories [1]: the interferometric methods using crystals [2,3], the propagation-based methods [4,5], the analyzer-based methods [6–8], the grating interferometric methods [9–12] and the grating non-interferometric methods [13–16]. In this work, the moiré imaging method [17–19] in combination with a Talbot–Lau imaging system is applied.

Using a Talbot–Lau interferometer [12,20–26], information about the phase-shift and the scattering properties of an object can be obtained simultaneously [12,27]. Therefore, this technique is very promising for applications in medical imaging [12,27–30] and non-destructive testing [12]. It is, for example, possible to depict micro calcifications in mammography for earlier diagnosis of breast tumors [28,29] or lung tissue to diagnose diseases like emphysema [30].

However, with regard to clinical applications, it is desirable to keep the imaging process as short as possible. In conventional Talbot–Lau imaging, a phase-stepping has to be performed [11] to resolve the subpixel information. This is a time-consuming process as the acquisition time is increased by taking several acquisitions at different grating positions. Hence, this imaging process is prone to mechanical instabilities and motion artifacts.

In recent years, some attempts have been made to overcome this problem. Miao et al. [31] performed an electromagnetic phase-stepping method to overcome the mechanical requirements of stepping the grating precisely. In this publication, we apply the moiré imaging method. Thus, no mechanical requirements for the phase-stepping are necessary and it is a single-shot approach. Hence, motion artifacts and image artifacts due to vibrations of the setup can be reduced. For applying the moiré imaging method, a conventional Talbot–Lau interferometer can be used. The interferometer has to be slightly detuned to be able to observe a high frequency moiré pattern that can still be resolved by the detector. Then, only a single-shot exposure is necessary to retrieve information about the attenuation, differential phase-contrast and dark-field property of the object simultaneously. The information can be extracted from the Fourier transformation of the moiré pattern that is deformed by the object [17–19,32,33]. Compared to the phase-stepping approach, in moiré imaging, the spatial resolution is slightly reduced [33,34].

Nevertheless, this technique is easy to implement and shows great advantages due to short exposure times. Additionally, this method is robust concerning vibrations of the setup as it is a single-shot method. Only the reproducibility between reference and object measurement has to be given.

In this publication, an advanced reconstruction method for moiré imaging is presented to improve image quality and to overcome the problem of an inhomogeneous moiré pattern over a large field of view (FoV).

2. Materials and Methods

2.1. Moiré Imaging Using a Talbot–Lau Setup

In order to measure the effect of objects on the phase, the amplitude and the offset of the spatially coherent wavefront, a Talbot–Lau interferometer can be used [35,36] (Figure 1). In the following, the general functionality of a Talbot–Lau interferometer is explained as it is also used to perform moiré imaging in this publication.

To separate the influence of the object on the incident wave from the influence of external impacts as, for example, imperfections of the gratings, an image without an object in the beam path is acquired. Subsequently, this image is called reference or free-field image. The reference information is marked by a sub-index "*ref*".

Illuminating a phase grating G_1 with a wavefront that has a sufficient degree of spatial coherence, self-images of the grating can be observed in so-called Talbot distances. This is called Talbot effect. An object between the source and the grating G_1 can refract the incident wavefront [20,26]. Passing though the object, the wavefront is speeded up. This leads to a phase-shift resulting in the observed refraction. The speedup depends on the material and is described by the complex index of refraction $n = 1 - \delta + i\beta$, where β describes the attenuation of X-rays and δ the phase-shift property within the material [37]. The refraction leads to a local shift of the Talbot intensity pattern downstream of G_1 perpendicular to the direction of propagation and to the grating bars. The small angle scattering, called dark-field information, can be observed in a reduced contrast of the Talbot intensity pattern.

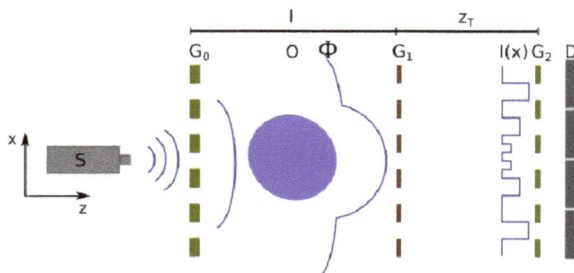

Figure 1. Setup of a Talbot–Lau interferometer. The object O deforms the X-ray wave front emitted by the source S. Because of the Lau and the Talbot effect (due to the gratings G_0 and G_1), a periodic intensity pattern can be observed in a Talbot distance z_T downstream of G_1. The period of the intensity pattern is small compared to the size of the detector pixel. In order to scan the pattern, a third grating G_2 has to be placed in a Talbot distance of G_1 in front of the detector. It is used either to sample the intensity pattern in conventional Talbot–Lau imaging or to generate a moiré pattern that can be resolved by the detector. The distance l can be be determined applying the intercept theorem.

The detector pixels are usually much larger than the period of the intensity pattern. Thus, the deformation of the self-image of G_1 can not be measured directly by the detector. Therefore, an analyzer grating G_2 has to be placed in a Talbot distance behind G_1. In the presented moiré imaging procedure, G_2 has to be slightly rotated around the beam axis compared to the self-image of G_1. For the ideal case, G_2 shows the same periodicity as the self-image of G_1 in the Talbot-distance. Thus, the superposition of both periodic structures that are rotated towards each other leads to a periodic pattern of lower frequency, called moiré pattern [38]. As the period of this moiré pattern is larger than the pixel size, it can be directly observed by the detector. Deformations of the self-image of G_1 caused by an object result in deformations of the moiré pattern.

As described, for example, by Takeda et al. [17] for the optical energy regime and Bennett et al. and Bevins et al. [18,19] for the X-ray regime, the information about attenuation, differential phase-contrast and dark-field can be extracted from the Fourier Transform (FT) of the measured moiré pattern image. According to Bennett et al. [18] and Bevins et al. [19], the mean intensity in each pixel can be reconstructed by taking the absolute value of the inverse FT of an area around the zero order harmonic $I = |\mathcal{F}_0^{-1}|$. The visibility can be calculated by taking the ratio of the absolute values of the inverse FT of the area around the first order harmonic and of the area around the zero order harmonic $V = \frac{|\mathcal{F}_1^{-1}|}{|\mathcal{F}_0^{-1}|}$. The phase information is reconstructed by taking the argument of the complex FT of the area around the first order harmonic $\phi = \arg(\mathcal{F}_1)$.

In this context, $\mathcal{F}_x$ is a pseudo code describing the area around the xth order harmonic of the Fourier transformation of the measured intensity image, and $\mathcal{F}_x^{-1}$ is the inverse Fourier transformation of this area. A more detailed explanation of how to determine the extent of these areas is given in Section 2.4.1. Furthermore, $|\cdot|$ means the absolute value of the complex Fourier transformation, and $\arg(\cdot)$ means the angle or the complex part of the complex Fourier transformation.

The three imaging observables can be calculated for each pixel with I, V and ϕ as follows:

$$\text{attenuation image:} \qquad T = -\ln\left(\frac{I_{0,\text{obj}}}{I_{0,\text{ref}}}\right), \qquad (1)$$

$$\text{differential phase-contrast image:} \qquad d_\phi = \left(\phi_{\text{obj}} - \phi_{\text{ref}}\right)(\text{mod } 2\pi), \qquad (2)$$

$$\text{dark-field image:} \qquad D = -\ln\left(\frac{V_{\text{obj}}}{V_{\text{ref}}}\right). \qquad (3)$$

Using a polychromatic X-ray source with a large focus, a third grating G_0 that is placed right behind the focus is required. It creates mutually independent slit sources whereof each of them behaves like a single coherent source [12,39,40].

The procedure to define "an area around the zero order harmonic" respectively "an area around the first order harmonic" is not unique. As an example, a sharp cut around the zero order harmonic of the moiré pattern eliminates high spatial frequencies and thus leads to blurred edges. Furthermore, by applying sharp cuts in the frequency domain of objects with sharp edges, ringing artifacts can be observed in the reconstructed images. In the following, an improved reconstruction technique will be introduced. Additionally, due to grating and setup imperfections, the moiré pattern is not regular over the whole FoV for a large setup. By reconstructing the whole matrix at once, the irregularities of the moiré pattern lead to artifacts in the reconstructed image. To avoid these artifacts, a piecewise reconstruction algorithm is presented in the following. This approach is subsequently called *sliding window reconstruction*. It is important to remark that this method addresses a different problem than the sliding window phase stepping presented by Zanette at al. [41].

2.2. Setup

The image data used in this publication are acquired with a Talbot–Lau interferometer. The setup is described in the following. A Siemens MEGALIX Cat Plus 125/40/90-125GW (Siemens, Munich, Germany) is used as an X-ray source. The detector is a Teledyne Dalsa Shad-o-Box (Teledyne DALSA, Waterloo, ON, Canada) with 50 µm pixelpitch. Tables 1 and 2 show further parameters of the setup.

Table 1. Distances between the components.

Components	Distance
source grating (G_0) - sample	124 cm
sample - phase grating (G_1)	10 cm
phase grating (G_1) - analyzer grating (G_2)	13 cm

Table 2. Parameters of the gratings.

	G0	G1	G2
material	Au	Au	Au
period (µm)	24.39	4.37	2.4
height of bars (µm)	180	6.4	90
duty cycle	0.5	0.75	0.5

All measurements are performed using a 40 kVp spectrum.

2.3. Measurement Samples

The phantom consists of three components (see Figure 2). The components are chosen in order to receive a constant signal in each image modality. The component on top of the image is a synthetic sponge of constant thickness causing a constant dark-field signal. In the middle, a step wedge made from Polymethyl methacrylate (PMMA). is used to examine the behavior of the attenuation image. The component on the bottom is a PMMA wedge of constant slope. This should result in a constant differential phase signal. Figures 3, 4, 6 and 7 show an image section comprising the edge of the sponge and of the step wedge marked by the green rectangle. Thus, the reconstruction details are more visible. Due to the high magnification of the images of the green box, these images seem to be pixilated. In these figures, the contrast is enhanced to visualize the artifacts and to emphasize the reconstruction details. Thus, the object information is saturated and there seem to be overexposed and underexposed areas in the images. To better detect the object information, a more appropriate choice of contrast is shown in Figure 9, which shows the images of the whole phantom. This area is marked by the red rectangle in Figure 2 and the corresponding images are shown in Figure 9.

The image of a human finger (ex-vivo) is shown in Figure 10. The specimen is provided by the Anatomical Institute II of Friedrich-Alexander University Erlangen-Nuremberg.

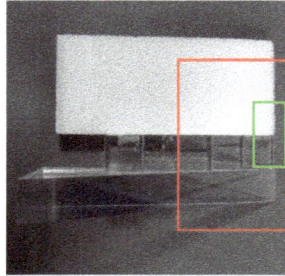

Figure 2. Photograph of the phantom. It consists of a synthetic sponge of constant thickness (**top**), a step wedge made from PMMA (**middle**) and a PMMA wedge with a constant slope (**bottom**). The red rectangle shows the area of the phantom that is imaged. The green rectangle marks the region that is imaged in the more detailed pictures.

2.4. Reconstruction Algorithm

2.4.1. Frequency Filters

Figure 3c shows the Talbot-moiré image obtained for the green marked area of the phantom shown in Figure 2. Figure 3a shows the corresponding reference image without phantom in the beam path. The moiré pattern generated by the superposition of the self-image of G_1 and the slightly rotated G_2 is clearly visible. Three peaks related to this pattern can be observed in the Fourier transformed image in Figure 3b of the reference image and in Figure 3d of the object image.

The central peak in Figure 3b,d, called the 0th harmonic (red arrow in Figure 3b,d, is located around the zero frequency position, which is in the center of the frequency space. This peak is not as dominant in the image of the FT of the reference image (Figure 3b) as in the image of the FT of the object image (Figure 3d). The reference image is dominated by the moiré pattern and no object information is contributing to the low-frequency signal like in the image of the object. The two remaining peaks, called the 1st and −1st harmonics (green arrows in Figure 3b,d, are located around the frequency of the moiré pattern ($\pm f_{moiré}$). The scattering and the phase-shift properties of the object are coded in a reduced contrast and a shift of the moiré pattern, respectively. Thus, the visibility and phase-shift information is represented in the 1st and −1st harmonics and their surroundings, whereas the attenuation information is coded in the area around the 0th harmonic. Consequently, as mentioned above and described for example by Bennett et al. [18] and Bevins et al. [19], the information of the harmonics has to be reconstructed separately. As the Fourier space is symmetric to the zero frequency, only positive frequencies are used to reconstruct the images.

If the different harmonics in the Fourier space are separated by a sharp cut, ringing artifacts and blurring will be observed in the reconstructed images (Figure 4). Ringing artifacts occur at sharp edges in the reconstructed images due to the fact that sharp edges are represented by a superposition of nearly all frequencies in Fourier space. By setting all high frequencies to zero, the edge is not represented correctly anymore. Ringing-like artifacts occur in the reconstructed images like shadows of the edge. These artifacts can be reduced if the different harmonics in Fourier space are not separated by a sharp cut. It is more appropriate to separate the harmonics by applying a Gaussian-shaped frequency filter centered at the maximum of the harmonic, which should be separated. It has to be considered that the Gaussian cut acts as a low pass filter. Thus, it has to be as sharp as possible to minimize the blurring effect. By contrast, it has to be as wide as possible to reduce the ringing artifacts. Hence, the width of the Gaussian is limited by half of the distance between the harmonics in Fourier space. The most feasible Gaussian has been found empirically. In addition, a region around the 1st harmonic is set to zero when cutting the 0th harmonic. Thus, no moiré artifacts are observed in the attenuation image.

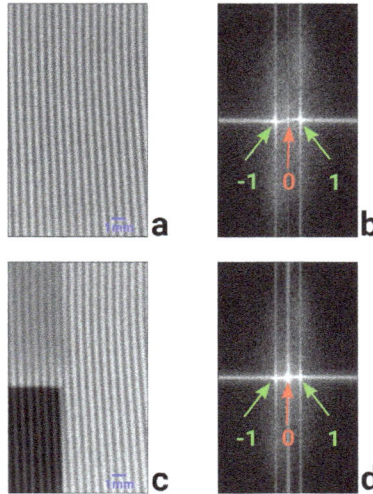

Figure 3. (a) reference Talbot-moiré image; (b) two-dimensional FT of (a); (c) selected area of the Talbot-moiré image of the phantom shown in Figure 2. The moiré pattern (stripes) is distorted by the sponge (**top**) and the PMMA step-wedge (**bottom**); (d) two-dimensional FT of (c). The green arrows mark the −1st and 1st harmonics that are located at the frequency of the moiré pattern. The red arrow marks the 0th harmonic that is located at the zero frequency position. The images correspond to the green area in Figure 2.

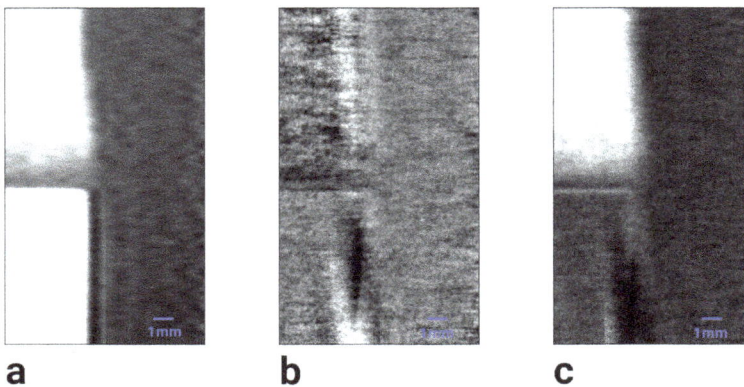

Figure 4. Attenuation (**a**), differential phase-contrast (**b**) and dark-field (**c**) image of the phantom's sponge (**top**) and the PMMA step-wedge (**bottom**). Ringing artifacts can be observed at the edges of the wedge. The images correspond to the green area in Figure 2.

Bennett et al. [18] and Bevins et al. [19] only consider a small area around the harmonics for reconstructing the different image domains. Neglecting the high frequencies causes blurring in the reconstructed images. For reconstructing the different image domains, it is necessary to suppress the harmonics that do not contribute to the image information. Nevertheless, it is not necessary to suppress high frequencies that are not correlated with the moiré pattern.

Thus, in this publication, improved cutting filters in Fourier space are presented in order to enhance image quality and avoid blurring artifacts.

The optimized filters are shown in Figure 5c,d in comparison to the sharp, conventionally used filters a,b. All cuts are performed by a Gaussian distribution whose standard deviation is as small as possible to avoid blurring artifacts without receiving ringing artifacts. In addition, in Figure 5c, the cut of the 0th harmonic towards the frequencies larger than the 1st harmonic is a very narrow Gaussian distribution. Furthermore, some values around the minimum of the Gaussian are set to zero as the smoothing effect of the low gradient of the Gaussian towards its extremum has to be reduced. Thus, it is difficult to recognize the Gaussian-like shape in the depiction.

Using these filters, it is ensured that only frequencies are set to zero, which carry interfering information for the single image domain. Especially when cutting the 0th harmonic, high frequencies are not neglected. This of course leads to a higher noise level in the reconstructed images. However, sharp edges are also less blurred as the high frequencies are necessary to depict edges as sharp as possible.

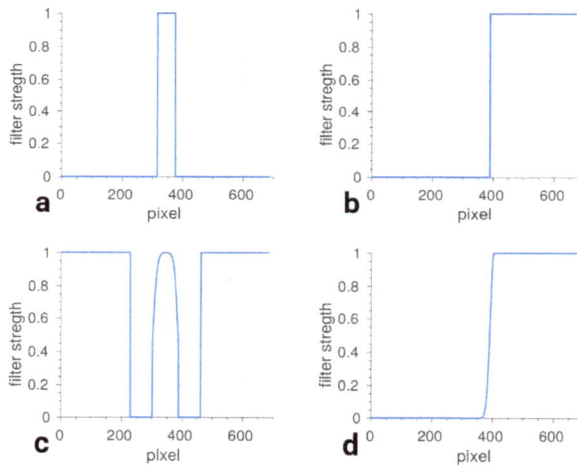

Figure 5. Sharp rectangular frequency filter to select the 0th harmonic (**a**) and 1st harmonic (**b**). Gaussian shaped filter to select the 0th harmonic (**c**) and 1st harmonic (**d**).

2.4.2. Sliding Window Reconstruction

The moiré pattern is not completely regular over a large FoV. This is, for example, due to grating imperfections and misalignments of the gratings. Hence, the frequency and the orientation of the moiré pattern vary slightly over the FoV. Reconstructing the images as described above a constant frequency and orientation is required. Otherwise, artifacts occur that can especially be seen in the attenuation image, e.g., in Figure 6a.

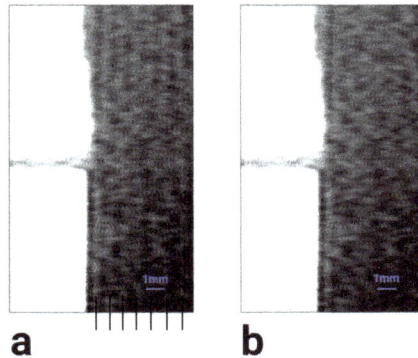

Figure 6. Attenuation images, reconstructed with a FT filter applied to the whole moiré image (**a**) and the sliding window reconstruction (**b**). Only the sponge (**top**) and the step-wedge (**bottom**) are shown. On the right of the one-step reconstruction, a moiré pattern is still visible (**a**). The fringes of the moiré pattern are marked by the black lines on the bottom of the image. These artifacts can be reduced strongly by performing the windowed reconstruction (**b**). The images correspond to the green area in Figure 2.

Therefore, it is preferable to limit the reconstructed areas to windows with a constant frequency of the moiré pattern. Additionally, the window should be small enough that the orientation of the pattern does not change within it. It has been empirically found that the smallest reconstruction window leading to meaningful results comprises two moiré periods. Thus, it is ensured that artifacts of defect detector pixels within one period of the moiré pattern or severe distortion of the pattern due to the object in the beam path do not prevent the algorithm to detect the correct superposed pattern. Nevertheless, the reconstructing window size is small enough that the period of the moiré pattern can be assumed as constant. Thus, for example, an image superposed by a moiré pattern with a period of 15 detector pixels can be reconstructed using a minimal window size of 30×1 pixels. The window should be orientated perpendicular to the moiré fringes to be able to analyze the fringes. Each window is reconstructed separately. Before Fourier transforming the window, each single window should be weighted with a Hanning filter [17] to avoid discontinuities at the margin of the windows. At the end, the windows are stitched together. Therefore, it is important to overlap the windows to avoid margin artifacts. However, at sharp edges, it is sometimes not possible to use the minimal window size to reconstruct the images because the moiré pattern is severely distorted by an edge of the object. Thus, the different harmonics cannot be detected and separated anymore. Hence, a compromise has to be made between a small reconstruction window to avoid artifacts due to inhomogeneities and a large reconstruction window in order to detect the moiré period even at edges in the image. This piecewise reconstruction technique is subsequently called *sliding window reconstruction*. In Figure 6b, it can be seen that, by using the sliding window reconstruction, artifacts caused by an inhomogenious moiré pattern can be reduced significantly.

Due to spatial discontinuities at the edge of the reconstruction window, artifacts can be observed at the margin of the reconstructed windows [17]. To avoid these artifacts, only the inner third of the reconstructed window is taken to stick the image together. Thus, the reconstruction windows are overlapping each other.

Using the above described techniques, it is possible to reconstruct images mostly artifact-free (Figure 7).

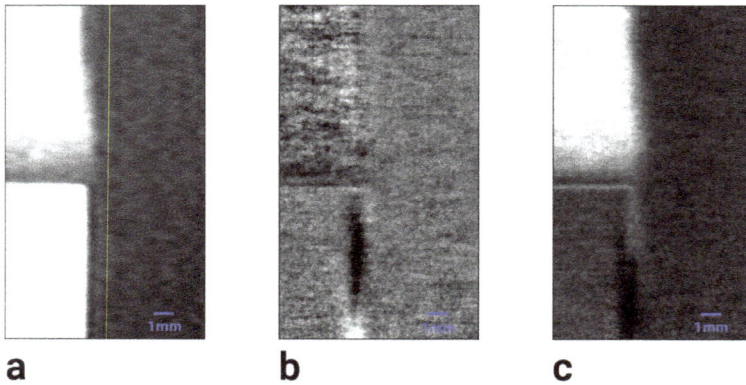

Figure 7. Attenuation (**a**), differential phase-contrast (**b**) and dark-field (**c**) image of the phantom reconstructed by using the Gaussian shaped filters and the sliding window reconstruction. The sponge (**top**) and the PMMA step-wedge (**bottom**) are shown. No artifacts are observed in the images. The images correspond to the green area in Figure 2.

3. Results

To evaluate the ability to reconstruct sharp edges by the different reconstruction methods, a lineplot of Figure 4a is compared to a lineplot of Figure 7a. The lineplot is taken from the edge of the step wedge on the bottom of the figure. It is shown in Figure 8. The blue line shows the edge reconstructed with a sharp cut in the frequency space. The red line shows the edge reconstructed with the Gaussian cut. Ringing of the blue line can be observed in front of and behind the edge. This ringing is reduced in the red line. Applying the sharp cut, high frequencies are set to zero, cutting the zero harmonic order (Figure 5a). This is not the case for the presented Gaussian cut (Figure 5c). Therefore, the blue line is smoothed compared to the red line. Furthermore, it can be observed that the red line shows a steeper gradient at the place of the edge than the blue line. Thus, the edge seems to be better reconstructed applying the Gaussian filter.

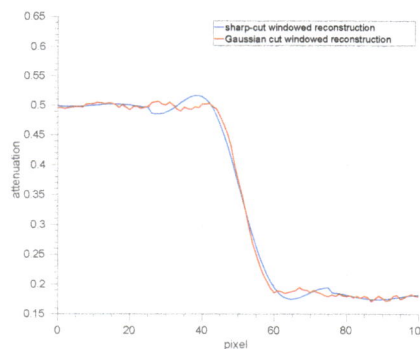

Figure 8. Lineplots of the edge of the step wedge in Figures 4a and 7a.

The contrast to noise ratios (CNRs) of the three image modalities have been evaluated for the moiré imaging approach in comparison to the conventional phase-stepping approach. The results of the measurement are shown in Table 3. The regions of interest (ROIs) that were used to calculate the CNRs are marked in Figure 9. It can be seen that the CNR values of the differential phase-contrast image and the dark-field image yield comparable results for both reconstruction methods. They are only slightly better for the phase-stepping approach. In the attenuation images, the moiré imaging approach leads to an even higher CNR value than the phase-stepping approach. This is due to the smoothing effect of the moiré imaging reconstruction of the attenuation image. Neglecting the frequencies around the first harmonics reduces the noise in the attenuation image. Reconstructing the dark-field image, all frequencies are taken into account as images gained with both filters are combined to calculate the dark-field image. For the differential phase-contrast image, only frequencies around the zero order harmonic are neglected. Thus, the smoothing in the dark-field and the differential phase-contrast image are not as emphasized as in the attenuation image.

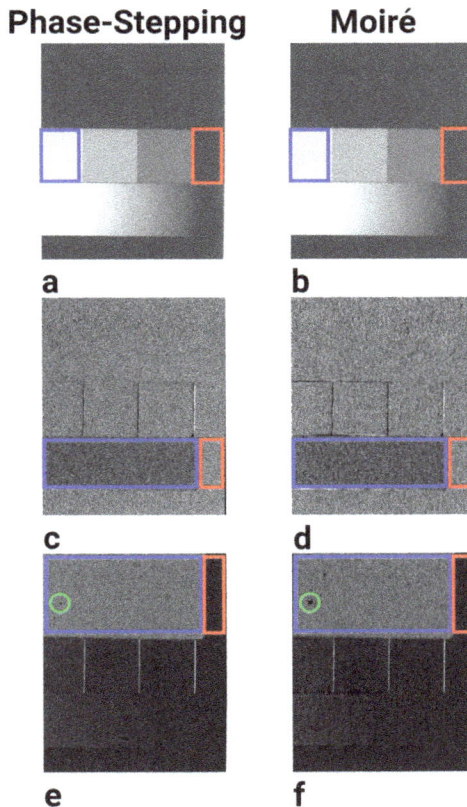

Figure 9. Images of the phantom shown in Figure 3 (red marked area). The ROIs that have been chosen to calculate the CNR values in Table 3 are shown. In the left column (**a**,**c**,**e**), the images are obtained using the phase-stepping approach, in the right column (**b**,**d**,**f**), the images are obtained using the moiré imaging approach. The attenuation images (**a**,**b**), the differential phase-contrast images (**c**,**d**) and the dark-field images (**e**,**f**) are shown. The green circle in the dark-field images (**e**,**f**) mark an artifact that is more emphasized in the moiré imaging approach.

Table 3. Contrast to noise ratios of the three image modalities for both imaging methods. All images are acquired with a dose of 900 mAs at a peak voltage of 40 kVp.

	Phase-Stepping	Moiré Imaging
attenuation	21.06 ± 0.05	25.91 ± 0.06
differential phase-contrast	1.14 ± 0.01	0.80 ± 0.01
dark-field	3.65 ± 0.01	2.84 ± 0.01

For further investigations, a human finger has been imaged to show the capabilities of the improved reconstruction algorithm. The three image modalities reconstructed with the conventional technique and with the improved algorithm, in comparison with the phase-stepping method, are shown in Figure 10. Especially in the differential phase-contrast and the dark-field image that are reconstructed with the conventional algorithm (Figure 10f,i) moiré artifacts can be seen. These artifacts can be completely removed applying the presented, improved reconstruction algorithm (Figure 10e,h). It can be seen that the image qualities of the phase-stepping approach and the moiré imaging approach reconstructed with the improved algorithm are comparable.

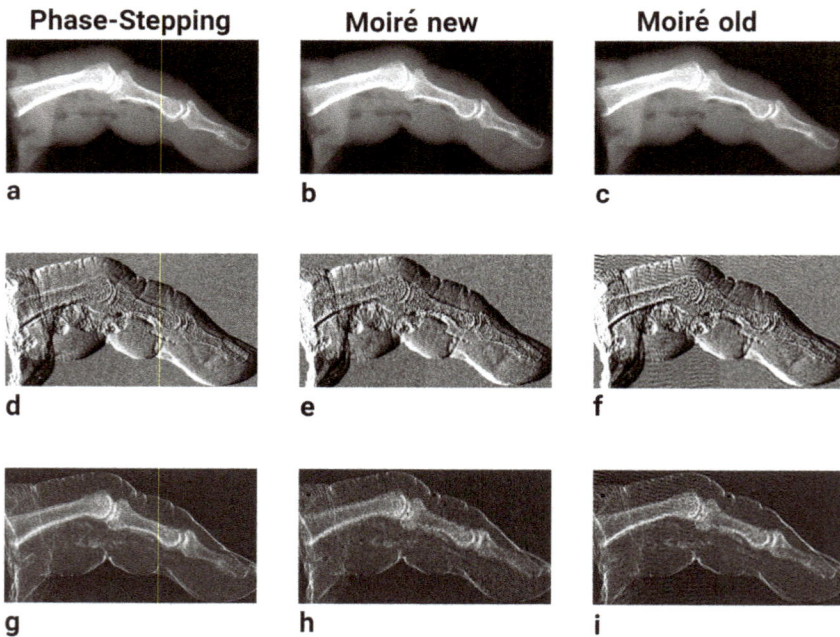

Figure 10. Images of a human finger, ex-vivo. The images in the left column (**a**,**d**,**g**) are reconstructed using the phase-stepping approach. In the middle (**b**,**e**,**h**), the images of the moiré imaging approach using the improved algorithm are depicted. In the right column (**c**,**f**,**i**), the images of the moiré approach reconstructed with the conventional method are shown. (**a**,**b**,**c**) attenuation images, (**d**,**e**,**f**) differential phase-contrast images and (**g**,**h**, **i**) dark-field images. The corresponding images are obtained with the same dose of 450 mAs at 40 kVp and are plotted with the same color scale.

4. Discussion

This work shows that it is possible to depict an object in attenuation, differential phase-contrast and dark-field image with the single-shot moiré X-ray imaging method. The CNR values of the images reconstructed with the moiré imaging approach are comparable to those of the phase-stepping images. Nevertheless, the spatial resolution is reduced in the images acquired by moiré imaging. This can be seen at the edges of the phantom components. The spatial resolution cannot be directly measured using a line phantom because the structures of the line phantom superpose with the structures of the moiré pattern. This leads to a further moiré pattern and thus the line phantom cannot be reconstructed correctly. In future experiments, it would be interesting to quantify the loss in spatial resolution by different approaches. This is a complex problem and needs further investigation.

It has to be mentioned that the object interested in medical imaging normally does not show periodic structures on the scale of the moiré pattern. Thus, the correct reconstruction of medical images should not be a problem. The sensitivity is the same as in the phase-stepping approach.

Furthermore, artifacts resulting from grating imperfections are more emphasized in moiré imaging. This can, for example, be seen in the dark-field image (Figure 9e,f) of the sponge (top), near the left margin of the image. An area with low signal (black dot) is visible for both image acquisition domains, but it is more emphasized in the moiré imaging. The area is marked by a green circle.

Nevertheless, with the help of the improved reconstruction technique, it is possible to reconstruct images with a good image quality as could be shown by comparing the CNR values. Thus, it seems appropriate to implement this single-shot imaging method in Talbot–Lau interferometry and improve its feasibility for workflow processes in clinical or industrial domains. Image acquisition time is mainly limited by the read-out time of the detector. As the moiré images are acquired by a single-shot approach, the detector is read out once for the moiré imaging approach. For the phase-stepping approach, the detector is read out after each phase-step. In the presented case, nine phase-steps have been performed. Additionally, in the phase-stepping approach, the time that is necessary to move the grating has to be added to the whole imaging time. To further accelerate the moiré imaging process, a high photon flux is advisable. Thus, the exposure time of the single-shot exposure can be shortened while keeping the dose constant. A higher photon flux cannot be used to accelerate the imaging process in the phase-stepping procedure. The exposure time of each phase-step is very short, so there is no need to shorten the exposure time for a single phase-step.

Hence, artifacts due to the object motion or setup instabilities can be reduced. For example, in lung imaging, the moiré imaging approach opens up new possibilities as motion artifacts due to breathing of the patient are no longer a problem. In particular, the moiré imaging approach is very promising regarding lung imaging due to its single-shot character. In this case, no high spatial resolution is necessary as especially the scattering properties of the lung tissue that are related to the air-filled aveola structure is of interest. Additionally, no high spatial resolution is necessary for depicting a lung in dark-field image. Hence, the reduced spatial resolution of moiré imaging is no problem for this application. There are many more fields of application for moiré imaging that should be evaluated in the future. For example, due to the fast image acquisition process, single-shot moiré imaging methods allow X-ray phase-contrast fluoroscopy or CT imaging. This is an important step towards the daily use of X-ray phase-contrast imaging in medical or industrial applications.

Author Contributions: Florian Horn, Veronika Ludwig, Georg Pelzer and Jens Rieger conceived and designed the experimental setup; Maria Seifert and Michael Gallersdörfer formulated research goals; Michael Gallersdörfer performed the experiments; M.S. analyzed the data; Maria Seifert and Max Schuster contributed analysis tools; Maria Seifert wrote the paper. Thilo Michel and Gisela Anton supervised the project.

Acknowledgments: We would like to acknowledge financial and technical support of this work by Siemens Healthcare GmbH. In addition, the authors would like to acknowledge the contribution of the Anatomical Institute II of the Friedrich-Alexander-University (Germany), who provided the specimen. Additionally, we would like to acknowledge the Karlsruhe Nano Micro Facility (KNMF), a Helmholtz Research Infrastructure at Karlsruhe Institute of Technology (KIT) and microworks GmbH for fabricating our gratings.

Conflicts of Interest: The authors declare no conflict of interest. The founding sponsors had no role in the design of the study; in the collection, analyses, or interpretation of data; in the writing of the manuscript, and in the decision to publish the results.

References

1. Diemoz, P.; Bravin, A.; Coan, P. Theoretical comparison of three X-ray phase-contrast imaging techniques: Propagation-based imaging, analyzer-based imaging and grating interferometry. *Opt. Express* **2012**, *20*, 2789–2805. [CrossRef] [PubMed]
2. Momose, A.; Takeda, T.; Itai, Y.; Hirano, K. Phase-contrast X-ray computed tomography for observing biological soft tissues. *Nat. Med.* **1996**, *2*, 473–475. [CrossRef] [PubMed]
3. Lewis, R.A.; Hall, C.J.; Hufton, A.P.; Evans, S.; Menk, R.H.; Arfelli, F.; Rigon, L.; Tromba, G.; Dance, D.R.; Ellis, I.O.; et al. X-ray refraction effects: Application to the imaging of biological tissues. *Br. J. Radiol.* **2003**, *76*, 301–308. [CrossRef] [PubMed]
4. Snigirev, A.; Snigireva, I.; Kohn, V.; Kuznetsov, S.; Schelokov, I. On the possibilities of X-ray phase contrast microimaging by coherent high-energy synchrotron radiation. *Rev. Sci. Instrum.* **1995**, *66*, 5486–5492. [CrossRef]
5. Wilkins, S.; Gureyev, T.; Gao, D.; Pogany, A.; Stevenson, A. Phase-contrast imaging using polychromatic hard X-rays. *Nature* **1996**, *384*, 335–338. [CrossRef]
6. Foerster, E.; Goetz, K.; Zaumseil, P. Double crystal diffractometry for the characterization of targets for laser fusion experiments. *Krist. Tech.* **1980**, *15*, 937–945. [CrossRef]
7. Chapman, D.; Thomlinson, W.; Johnston, R.; Washburn, D.; Pisano, E.; Gmuer, N.; Zhong, Z.; Menk, R.; Arfelli, F.; Sayers, D. Diffraction enhanced X-ray imaging. *Phys. Med. Biol.* **1997**, *42*, 2015–2025. [CrossRef] [PubMed]
8. Bravin, A. Exploiting the X-ray refraction contrast with an analyser: The state of the art. *J. Phys. D Appl. Phys.* **2003**, *36*, A24–A29. [CrossRef]
9. David, C.; Noehammer, B.; Solak, H.; Ziegler, E. Differential X-ray phase contrast imaging using a shearing interferometer. *Appl. Phys. Lett.* **2002**, *81*, 3287–3289. [CrossRef]
10. Momose, A.; Kawamoto, S.; Koyama, I.; Hamaishi, Y.; Takai, K.; Suzuki, Y. Demonstration of X-Ray Talbot Interferometry. *Jpn. J. Appl. Phys.* **2003**, *52*, L866–L868. [CrossRef]
11. Weitkamp, T.; Diaz, A.; David, C.; Pfeiffer, F.; Stampanoni, M.; Cloetens, P.; Ziegler, E. X-ray phase imaging with a grating interferometer. *Jpn. J. Appl. Phys.* **2005**, *13*, 6296–6304. [CrossRef]
12. Pfeiffer, F.; Weitkamp, T.; Bunk, O.; David, C. Phase retrieval and differential phase-contrast imaging with low-brilliance X-ray sources. *Nat. Phys.* **2006**, *2*, 258–261. [CrossRef]
13. Olivo, A.; Arfelli, F.; Cantatore, G.; Longo, R.; Menk, R.; Pani, S.; Prest, M.; Poropat, P.; Rigon, L.; Tromba, G.; et al. An innovative digital imaging set-up allowing a low-dose approach to phase contrast applications in the medical field. *Med. Phys.* **2001**, *28*, 1610–1619. [CrossRef] [PubMed]
14. Olivo, A.; Speller, R. A coded-aperture technique allowing X-ray phase contrast imaging with conventional sources. *Appl. Phys. Lett.* **2007**, *91*, 1–3. [CrossRef]
15. Olivo, A.; Ignatyev, K.; Munro, P.R.T.; Speller, R.D. Noninterferometric phase-contrast images obtained with incoherent X-ray sources. *Appl. Opt.* **2011**, *50*, 1765–1769. [CrossRef] [PubMed]
16. Endrizzi, M.; Basta, D.; Olivo, A. Laboratory-based X-ray phase-contrast imaging with misaligned optical elements. *Appl. Phys. Lett.* **2015**, *107*, 124103. [CrossRef]
17. Takeda, M.; Ina, H.; Kobayashi, S. Fourier-transform method of fringe-pattern analysis for computer-based topography and interferometry. *J. Opt. Soc. Am.* **1982**, *72*, 156–160. [CrossRef]
18. Bennett, E.E.; Kopace, R.; Stein, A.F.; Wen, H. A grating-based single-shot X-ray phase contrast and diffraction method for in vivo imaging. *Med. Phys.* **2010**, *37*, 6047–6054. [CrossRef] [PubMed]
19. Bevins, N.; Zambelli, J.; Li, K.; Qi, Z.; Chen, G.H. Multicontrast X-ray computed tomography imaging using Talbot–Lau interferometry without phase stepping. *Med. Phys.* **2012**, *39*, 424–428. [CrossRef] [PubMed]
20. Talbot, H. Facts relating to optical science. *Opt. Commun.* **1836**, *9*, 401–407.
21. Lau, E. Beugungserscheinungen an Doppelrastern. *Ann. Phys.* **1948**, *437*, 417–423. [CrossRef]
22. Jahns, J.; Lohmann, A. The Lau effect (a diffraction experiment with incoherent illumination). *Opt. Commun.* **1979**, *28*, 263–267. [CrossRef]
23. Patorski, K. Incoherent Superposition of Multiple Self-imaging Lau Effect and Moiré Fringe Explanation. *Opt. Act Int. J. Opt.* **1983**, *30*, 745–758. [CrossRef]
24. Bartelt, H.; Jahns, J. Interferometry based on the Lau effect. *Opt. Commun.* **1979**, *30*, 268–274. [CrossRef]

25. Clauser, J.; Reinsch, M. New theoretical and experimental results in fresnel optics with applications to matter-wave and X-ray interferometry. *Appl. Phys. B Photophys. Laser Chem.* **1992**, *54*, 380–395. [CrossRef]

26. Suleski, T.J. Generation of Lohmann images from binary-phase Talbot array illuminators. *Appl. Opt.* **1997**, *36*, 4686–4691. [CrossRef] [PubMed]

27. Li, K.; Ge, Y.; Garrett, J.; Bevins, N.; Zambelli, J.; Chen, G.H. Grating-based phase contrast tomosynthesis imaging: Proof-of-concept experimental studies. *Med. Phys.* **2014**, *41*. [CrossRef] [PubMed]

28. Stampanoni, M.; Wang, Z.; Thüring, T.; David, C.; Roessl, E.; Trippel, M.; Kubik-Huch, R.A.; Singer, G.; Hohl, M.K.; Hauser, N. The first analysis and clinical evaluation of native breast tissue using differential phase-contrast mammography. *Investig. Radiol.* **2011**, *46*, 801–806. [CrossRef] [PubMed]

29. Michel, T.; Rieger, J.; Anton, G.; Bayer, F.; Beckmann, M.W.; Durst, J.; Fasching, P.A.; Haas, W.; Hartmann, A.; Pelzer, G.; et al. On a dark-field signal generated by micrometer-sized calcifications in phase-contrast mammography. *Phys. Med. Biol.* **2013**, *58*, 2713–2732. [CrossRef] [PubMed]

30. Bech, M.; Tapfer, A.; Velroyen, A.; Yaroshenko, B.; Pauwels, B.; Hostens, J.; Bruyndonckx, P.; Sasov, A.; Pfeiffer, F. In-vivo dark-field and phase-contrast X-ray imaging. *Sci. Rep.* **2013**, *3*. [CrossRef] [PubMed]

31. Miao, H.; Chen, L.; Bennett, E.E.; Adamo, N.M.; Gomella, A.A.; DeLuca, A.M.; Patel, A.; Morgan, N.Y.; Wen, H. Motionless phase stepping in X-ray phase contrast imaging with a compact source. *Proc. Natl. Acad. Sci. USA* **2013**, *110*, 19268–19272. [CrossRef] [PubMed]

32. Weitkamp, T.; Nöhammer, B.; Diaz, A.; David, C.; Ziegler, E. X-ray wavefront analysis and optics characterization with a grating interferometer. *Appl. Phys. Lett.* **2005**, *86*, 054101. [CrossRef]

33. Momose, A.; Yashiro, W.; Maikusa, H.; Takeda, Y. High-speed X-ray phase imaging and X-ray phase tomography with Talbot interferometer and white synchrotron radiation. *Opt. Express* **2009**, *17*, 12540–12545. [CrossRef] [PubMed]

34. Momose, A.; Yashiro, W.; Harasse, S.; Kuwabara, H. Four-dimensional X-ray phase tomography with Talbot interferometry and white synchrotron radiation: Dynamic observation of a living worm. *Opt. Express* **2011**, *19*, 8423–8432. [CrossRef] [PubMed]

35. Lohmann, A.; Silva, D. An interferometer based on the Talbot effect. *Philos. Mag. Ser. 3* **1971**, *2*, 413–415. [CrossRef]

36. Pfeiffer, F.; Weitkamp, T.; David, C. X-ray phase contrast imaging using a grating interferometer. *Europhys. News* **2006**, *37*, 13–15. [CrossRef]

37. Fitzgerald, R. Phase-sensitive X-ray imaging. *Phys. Today* **2000**, *53*, 23–26. [CrossRef]

38. Amidror, I. *The Theory of the Moiré Phenomenon-Volume I: Periodic Layers*; Springer: London, UK, 2009.

39. David, C.; Pfeiffer, F.; Weitkamp, T. Interferometer for Quantitative Phase Contrast Imaging and Tomography with an Incoherent Polychromatic X-ray Source. European Patent Application No. EP05012121, 13 December 2006.

40. Weitkamp, T.; David, C.; Kottler, C.; Bunk, O.; Pfeiffer, F. Tomography with grating interferometers at low-brilliance sources. In Proceedings of the Developments in X-Ray Tomography V, San Diego, CA, USA, 28 August 2016; Volume 6318, doi:10.1117/12.683851.

41. Zanette, I.; Bech, M.; Rack, A.; Le Duc, G.; Tafforeau, P.; David, C.; Mohr, J.; Pfeiffer, F.; Weitkamp, T. Trimodal low-dose X-ray tomography. *Proc. Natl. Acad. Sci. USA* **2012**, *109*, 10199–10204. [CrossRef] [PubMed]

Journal of
Imaging

MDPI

Article

Applications of Laboratory-Based Phase-Contrast Imaging Using Speckle Tracking Technique towards High Energy X-Rays

Tunhe Zhou [1,*], Fei Yang [2,†], Rolf Kaufmann [2] and Hongchang Wang [1]

1   Diamond Light Source, Harwell Science and Innovation Campus, Didcot OX11 0DE, Oxfordshire, UK;
    hongchang.wang@diamond.ac.uk
2   Empa, Swiss Federal Laboratories for Materials Science and Technology, 8600 Dübendorf, Switzerland;
    fei.yang@excillum.com (F.Y.); rolf.kaufmann@empa.ch (R.K.)
*   Correspondence: tunhe.zhou@diamond.ac.uk; Tel.: +44-1235-77-8017
†   Now at Excillum AB, Torshamnsgatan 35, 164 40 Kista, Sweden.

Received: 15 March 2018; Accepted: 8 May 2018; Published: 11 May 2018

Abstract: The recently developed speckle-based technique is a promising candidate for laboratory-based X-ray phase-contrast imaging due to its compatibility with polychromatic X-rays, multi-modality and flexibility. Previously, successful implementations of the method on laboratory systems have been shown mostly with energies less than 20 keV on samples with materials like soft tissues or polymer. Higher energy X-rays are needed for penetrating materials with a higher atomic number or that are thicker in size. A first demonstration using high energy X-rays was recently given. Here, we present more potential application examples, i.e., a multi-contrast imaging of an IC chip and a phase tomography of a mortar sample, at an average photon energy of 40 keV using a laboratory X-ray tube. We believe the results demonstrate the applicability of this technique in a wide range of fields for non-destructive examination in industry and material science.

Keywords: X-ray imaging; phase contrast; speckle; multi-contrast; tomography; chip; cement

1. Introduction

High energy X-rays are used for many applications in non-destructive examinations in industry or security for materials with high density or large thickness, because of their higher penetration ability. Biomedical imaging, such as dental and chest radiography, or large animal imaging, is also a field where high energy X-rays are applied. Research shows that even low-absorbing tissues, such as breast tissue, with the aid of phase-contrast imaging technique, can be imaged by X-rays with higher energies, providing the benefit of lower dose deposition [1].

X-ray phase-contrast imaging (XPCI) has drawn increasing attention as being complimentary or an alternative to the conventional attenuation-contrast method. From the phase contrast, the electron density of the material can be retrieved [2]. With the contrasts from both attenuation and phase, the complex refractive index of a material can be obtained. This allows the effective atomic number to be calculated and hence the quantitative characterization of the material [3]. Moreover, comparisons have been done theoretically and experimentally for the two contrast mechanisms and have shown that the phase contrast can provide higher signal-to-noise ratio for high spatial resolution imaging of materials within certain energy ranges [4–6].

While the attenuation of the X-ray intensity can be measured directly with an X-ray detector, the phase shift needs to be converted to a detectable signal and retrieved using post processing. Varieties of XPCI techniques have been developed during recent decades [7,8]. Some of them have a high demand on the temporal coherence of X-ray beams, which are more practical for synchrotron sources, such as crystal interferometry [9], and analyzer-based imaging, also known as diffraction-enhanced imaging [10]. For XPCI to be practical and more accessible for medical imaging and industrial non-destructive examination, laboratory sources are the preferred options. Several XPCI methods are compatible with polychromatic X-rays, namely propagation-, grating-, speckle-based imaging and edge illumination [11–14]. With the aid of monochromators, the analyzer-based technique has been realized using laboratory sources [15]. Energy-sensitive detectors can potentially enhance the imaging signal-to-noise ratio [16] and loosen a phase-contrast technique's demand on monochromaticity of a beam.

The speckle-based imaging (SBI) technique has been developed in recent years, starting at synchrotron facilities [17,18]. SBI has then been implemented on a laboratory-based liquid-metal jet source with an average X-ray energy less than 20 keV [13]. The imaged samples had materials with a low atomic number, such as organic polymers. A first demonstration of implementing SBI on high-energy X-ray laboratory source has been presented recently with an average energy over 100 keV [19]. For many XPCI methods, the challenge of being implemented to a high energy range involves the fabrication of corresponding optics. For example, an absorption grating with an aspect ratio of 143 was fabricated for the grating-based phase-contrast imaging technique for 100 keV [20]. The fabrication process can be both cost- and time-consuming. In comparison, steel wool was used as a random wavefront mask for SBI [19]. No stringent requirement on optics allows SBI to be easily and widely implemented. Here, we further investigate the applicability of SBI on commercially available laboratory sources towards high energies of the multi-contrast modalities and phase-contrast tomography. Tests from two samples are presented here: An integrated circuit (IC) chip and a cement mortar specimen, as examples of potential applications.

2. Materials and Methods

The experiments were conducted at Empa's Center for X-ray Analytics. The alignment is shown in Figure 1a. The source was a Hamamatsu source in an EasyTom system from Rx Solutions (Rx Solutions SAS, Chavanod, France), with a tungsten target operated at 80 kV with an emission current of 74 μA. Estimated from a simulated spectrum [21], the average photon energy was about 40 keV. The detector, located 1.03 m downstream of the source, was a CCD camera based on a scintillator coupled by a fibre optic plate. The camera was binned 2×2 during experiments, resulting in a pixel size of 18 μm. The camera sensor area is 36×24 mm^2, meaning the numerical aperture of the imaging system was about 35 mrad horizontally and 23 mrad vertically. The speckle pattern was generated by a stack of three layers of sandpaper (P220) with an average grit size of 68 μm [22], as shown in Figure 1b. Steel wool (Grade 0000) with an average fiber thickness of about 25 μm [23] packed to around 3 mm thick was also tested as wavefront modulator and is shown in Figure 1c. The speckle pattern from steel wool had higher visibility, but much higher absorption. If visibility is defined as the ratio of standard deviation and mean intensity within a subset window $V = \sigma/\bar{I}$, the average visibility of the speckle pattern from sandpaper is about 5% and from steel wool 17%. A subset window of 50×50 pixel^2 and an average over the image was used for calculation of visibility to provide better statistical data. However, the mean intensity in the speckle image from the sandpaper is 8 times higher than the image from the steel wool. Due to the limitations of the flux, the sandpaper was used for the experiments. The samples were mounted on a rotational stage located 0.2 m downstream of the source. The diffuser was mounted 7 cm away from the sample. The distances between each component in the alignment are flexible, which is easy to implement in any laboratory, compared to other phase-contrast imaging techniques, e.g., grating interferometry, which needs to match the grating periods and corresponding Talbot distances. With longer effective propagation distance, the speckle visibility is higher. A longer propagation distance also gives higher angular sensitivity, with the cost of lower flux density. A larger magnification means a smaller field of view. Therefore, there are tradeoffs for any alignment geometry and an optimization can only be discussed with a determined goal.

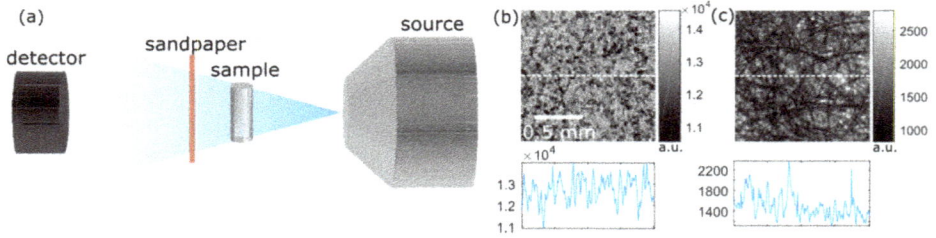

Figure 1. (**a**) Illustration of the experimental alignment. X-ray image of (**b**) sandpaper and (**c**) steel wool, and profiles from the dashed lines.

The principle of the speckle-tracking technique applied in the experiment is that the speckle patterns generated by the sandpaper function as a wavefront marker. The distortion of the wavefront can be retrieved by tracking the spatial change of the wavefront marker. Under the assumption that the phase is not changing rapidly, a rigid shift (u_x, u_y) of the speckle pattern is approximated. The wavefront refraction can hence be measured as $(\alpha_x, \alpha_y) = \frac{p}{z}(u_x, u_y)$, where p is pixel size and z is the propagation distance. Taking two speckle images with and without a sample in the beam, the shift of the speckle pattern between the two images can be calculated pixel-wise by normalized cross-correlation and then by polynomial fitting to achieve sub-pixel resolution [24]. The precision of the algorithm is 0.001 pixel, meaning that the limit of spatial and angular resolution in this experimental alignment are 18 nm and 24 nrad, respectively. The observable resolution critically depends on the noise level in the result, which is largely determined by the photon noise and visibility of the speckle pattern [25]. Under small angle assumption, the differential phase contrast is approximately proportional to the refraction angle as $\left(\frac{\partial \Phi}{\partial x}, \frac{\partial \Phi}{\partial y}\right) = \frac{2\pi}{\lambda}(\alpha_x, \alpha_y)$. The phase map can then be integrated from the phase gradients in two directions. Simultaneously, the transmission image T can be achieved by calculating the intensity change between the images; while the change of the visibility of the speckle patterns can be defined as the dark-field images $D = \frac{\sigma_{\mathrm{samp}}}{I_{\mathrm{samp}}} / \frac{\sigma_{\mathrm{ref}}}{I_{\mathrm{ref}}}$, providing information about the small angle scattering of a sample [26].

3. Results

The first result presented here is the multi-contrast radiography of an IC chip. The image was acquired with an exposure time of 5 min. The imaging processing uses a cross-correlation window width of 27 pixels with hamming weighting. The choice of the subset window size is a tradeoff between the spatial resolution and the noise level [25,27]. It is flexible, with a thumb of rule of being larger than the speckle size. Figure 2 shows the results of the multi-contrast images with the refraction angle in horizontal and vertical directions shown in Figure 2a,b, respectively; the phase shift in Figure 2c, the dark-field image in Figure 2d and the transmission image in Figure 2e. Figure 2f,g are the marked areas from the refraction angle Figure 2b and the vertical differential of the transmission image Figure 2e, respectively. The differentiation is taken to have a more visually straightforward comparison with the differential of the phase signal, as the derivative of a signal has the potential of higher sensitivity to small structures. From Figure 2f, we can distinguish structures beneath the resistor, such as the example indicated by the arrows in the figure, but in Figure 2e,g, there is not enough attenuation-contrast for the structures. For a non-destructive examination of strong absorbing products, like in this example, the speckle-based technique can realize multi-contrast imaging from a single data set and provide complimentary information. Structures that cannot be identified from the conventional attenuation signal can hence be examined from signals based on other modalities. The flexible experimental arrangement and low-cost wavefront modulator can allow this technique to be easily implemented for online quality control or production monitoring in laboratories or industry.

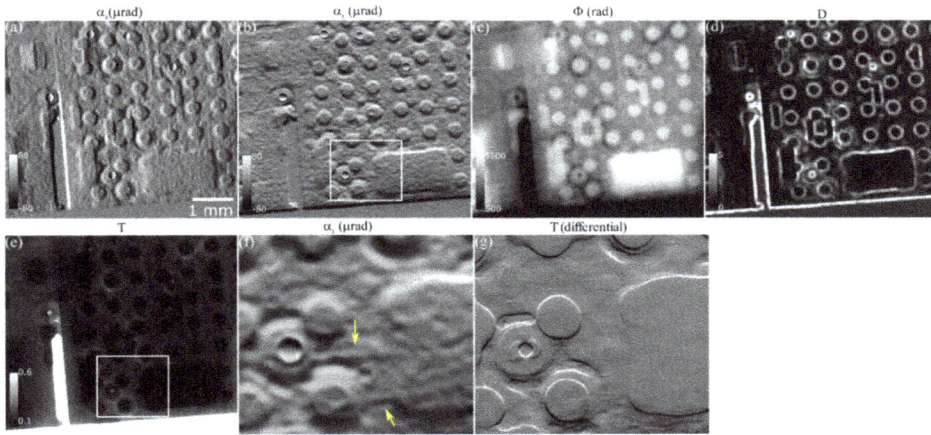

Figure 2. The multimodal images of a chip retrieved from the speckle-based imaging method: (**a**) refraction angle in horizontal and (**b**) in vertical direction, (**c**) phase shift (**d**) dark-field signal and (**e**) transmission image. (**f**) shows the refraction angle in vertical direction of the zoomed region in (**b**), and (**g**) is the transmission signal differentiated in vertical direction from the region of interest from (**e**).

The second example is a tomography scan of a mortar specimen. Cement mortar is one of the most commonly used construction materials. To understand the materials' physical properties, such as strength, permeability, etc., extensive knowledge of their microstructure and composition is essential. Various imaging techniques have been applied to study the microstructure, such as light microscopy, electron microscopy, etc. [28,29]. Being non-destructive, X-ray tomography has the advantage of conserving the original state of the sample and removes the need for sophisticated sample preparation [30,31]. Attenuation contrast works well between pores and groundmass, but faces challenges distinguishing other components. XPCI has been applied on porous samples using the grating-based technique at synchrotron facility and with a conventional X-ray tube [32,33]. High resolution phase-contrast tomography has been achieved using methods such as the Zernike phase-contrast method and ptychography, at synchrotron sources [34,35]. A complex refractive index was retrieved from the phase and absorption signal. With similarities in imaging technique to the grating-based method, the speckle-based method has the potential for a similar achievement, with a more cost-efficient and more flexible manner that, offers complimentary information for laboratory non-invasive examinations.

The mortar specimen was casted following the mixing design as described below: 0.51 water-to-cement ratio by mass, 50% volumetric fraction of sand (0.3–0.4 mm average size and 1 mm maximum size) and 19% volumetric fraction of cement CEM I 52.5 N (3.13 g/cm^3 density and 3440 cm^2/g Blaine fineness). After being cured for 91 days, a cylindrical mortar specimen was cored out of the large specimen, while being flushed with deionized water. The sample was dried in oven at 50 °C for 5 days, and then its lateral surface was covered with a 70-μm-thick polyimide film. Before the scan, the sample was kept in a dried state.

The scan was taken with 1 min exposure time for each projection and 992 projections over 360°. The width of the cross-correlation window was 33 pixels. Tomographic reconstruction of the phase images was done with Feldkamp-Davis-Kress (FDK) algorithm [36] with ram filter in Octopus software [37]. One reconstructed slice from phase data is shown in Figure 3a and the volume rendering in Figure 3b. As marked in Figure 3, from the reconstruction result we can clearly identify the different components in the sample. The brighter large particles are the sands, the brightest small particles are the unhydrated cement, the relative darker ground regions are the cement paste, and the darkest regions are the pores. The difference of the components is also shown in the line profile that passes through regions of cement paste, sand and cement, in the inset of Figure 3a. Some artefacts observed at the outer part of the sample as bright edges were induced by the strong edge enhancements due to the rapid phase change at the edge of the sample. It has been shown that these artefacts can be reduced by using the images with sample in the beam instead of the flatfield image in the processing algorithm to remove the edge enhancements [38]. The oscillation observed in the line profile in the cement paste and sand are partially due to the uncorrected ring artifact and the pores that are unresolved by the imaging system presented. The cement paste region, which has higher level gray value oscillation, in particular contains the pores that have a size of nm to tens of μm, which are far below the resolution of imaging system, but may shift the gray value. In addition, there might be unhydrated cement particles, which are below the level of spatial resolution. Due to these artifacts, the result presented here may not be highly quantitatively accurate, however it demonstrates the potential of applying the speckle-based imaging on tomography on a laboratory source towards high energies with future improvement.

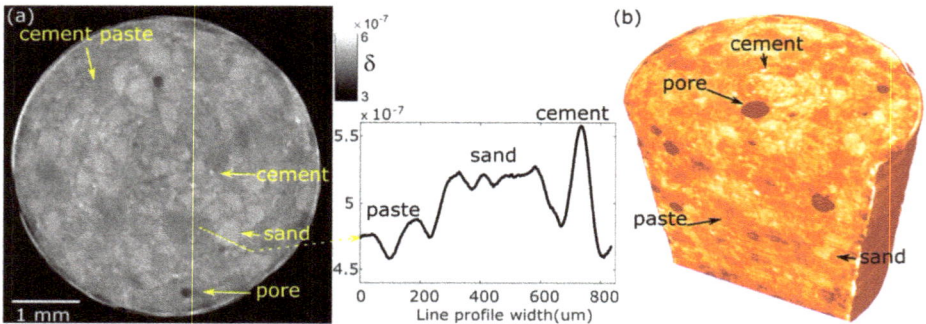

Figure 3. (**a**) Reconstructed tomographic slice of phase image of a mortar specimen. Inset shows the intensity profile of the dashed line marked in the slice. (**b**) Volume rendering from the reconstructed tomography.

Compared to previous study from Prade et al. [33], the source used in this study had over 10 times less power and effective pixel size that was about 30 times smaller. A longer exposure time was therefore needed. As rapid developments occur in X-ray tube research, there are already microfocus sources with higher power on market than the conventional tube used in Ref. [33], such as the metal-jet source. With a more powerful source, a more efficient detector, and without the absorption of the source and analyzer gratings as in the grating-based method, the speckle-based technique has the potential to have comparable or even shorter exposure time than in Ref. [33]. The spatial resolution for speckle tracking technique here is largely limited by the correlation window size but can be largely improved by using extra stages to scan the wavefront modulator, e.g., in 1D, 2D raster or a non-gird scheme [26,39,40]. The speckle size and visibility also affect the spatial resolution and noise level [24,25], which can be adjusted by changing to different diffusers for other applications.

4. Conclusions

The speckle-based imaging has been implemented on a laboratory source with an average energy of 40 keV. The results of the multi-contrast signal of a chip show structures which otherwise cannot be seen by conventional attenuation-contrast imaging. A first demonstration of a tomography of a cement mortar sample on a laboratory system using the speckle-based technique towards high energies has been presented. It shows the applicability of the technique for volumetric reconstructing of porous and composite materials. As the image quality can be further improved by utilizing a higher-power source,

J. Imaging **2018**, *4*, 69

more efficient detector, scanning scheme, etc., the speckle-based technique shows potential to be widely applied in the future for non-destructive imaging for industry and research in material science.

Author Contributions: T.Z. and F.Y. conceived, designed and performed the experiments; R.K. assisted conducting the experiments; T.Z. and F.Y. analyzed the data; T.Z., H.W., F.Y. and R.K. contributed the formation of the paper.

Acknowledgments: We acknowledge the financial support of the Swiss National Science Foundation (project numbers 143782 and 162572) and of the Helmholtz Virtual Institute for New X-ray analytic Methods in Materials Science (VI-NXMM). Part of this work was performed by the use of the Empa Platform for Image Analysis, http://empa.ch/web/s499/software-/-imaging-platform, at Empa's Center for X-ray Analytics. We would like to thank Carmelo Di Bella for the help in the preparation of the mortar specimens. We acknowledge Pietro Lura and Michele Griffa for the support of the project and help in manuscript preparation.

References

1. Diemoz, P.C.; Bravin, A.; Sztrokay-Gaul, A.; Ruat, M.; Grandl, S.; Mayr, D.; Auweter, S.; Mittone, A.; Brun, E.; Ponchut, C.; et al. A method for high-energy, low-dose mammography using edge illumination X-ray phase-contrast imaging. *Phys. Med. Biol.* **2016**, *61*, 8750–8761. [CrossRef] [PubMed]
2. Als-Nielsen, J.; McMorrow, D. *Elements of Modern X-ray Physics*; John Wiley and Sons: Hoboken, NJ, USA, 2001.
3. Zhihua, Q.; Joseph, Z.; Nicholas, B.; Guang-Hong, C. Quantitative imaging of electron density and effective atomic number using phase contrast ct. *Phys. Med. Biol.* **2010**, *55*, 2669.
4. Raupach, R.; Flohr, T. Performance evaluation of X-ray differential phase contrast computed tomography (PCT) with respect to medical imaging. *Med. Phys.* **2012**, *39*, 4761–4774. [CrossRef] [PubMed]
5. Lundstrom, U.; Larsson, D.H.; Burvall, A.; Takman, P.A.C.; Scott, L.; Brismar, H.; Hertz, H.M. X-ray phase contrast for CO_2 microangiography. *Phys. Med. Biol.* **2012**, *57*, 2603. [CrossRef] [PubMed]
6. Zambelli, J.; Bevins, N.; Qi, Z.H.; Chen, G.H. Radiation dose efficiency comparison between differential phase contrast CT and conventional absorption CT. *Med. Phys.* **2010**, *37*, 2473–2479. [CrossRef] [PubMed]
7. Momose, A. Recent advances in X-ray phase imaging. *Jpn. J. Appl. Phys.* **2005**, *44*, 6355–6367. [CrossRef]
8. Endrizzi, M. X-ray phase-contrast imaging. *Nucl. Instrum. Methods Phys. Res. A* **2018**, *878*, 88–98. [CrossRef]
9. Bonse, U.; Hart, M. An X-ray interferometer. *Appl. Phys. Lett.* **1965**, *6*, 155–156. [CrossRef]
10. Chapman, D.; Thomlinson, W.; Johnston, R.E.; Washburn, D.; Pisano, E.; Gmur, N.; Zhong, Z.; Menk, R.; Arfelli, F.; Sayers, D. Diffraction enhanced X-ray imaging. *Phys. Med. Biol.* **1997**, *42*, 2015–2025. [CrossRef] [PubMed]
11. Wilkins, S.W.; Gureyev, T.E.; Gao, D.; Pogany, A.; Stevenson, A.W. Phase-contrast imaging using polychromatic hard X-rays. *Nature* **1996**, *384*, 335–338. [CrossRef]
12. Pfeiffer, F.; Weitkamp, T.; Bunk, O.; David, C. Phase retrieval and differential phase-contrast imaging with low-brilliance X-ray sources. *Nat. Phys.* **2006**, *2*, 258–261. [CrossRef]
13. Zanette, I.; Zhou, T.; Burvall, A.; Lundström, U.; Larsson, D.H.; Zdora, M.; Thibault, P.; Pfeiffer, F.; Hertz, H.M. Speckle-based X-ray phase-contrast and dark-field imaging with a laboratory source. *Phys. Rev. Lett.* **2014**, *112*, 253903. [CrossRef] [PubMed]
14. Olivo, A.; Speller, R. A coded-aperture technique allowing X-ray phase contrast imaging with conventional sources. *Appl. Phys. Lett.* **2007**, *91*, 074106. [CrossRef]
15. Parham, C.; Zhong, Z.; Connor, D.M.; Chapman, L.D.; Pisano, E.D. Design and implementation of a compact low-dose diffraction enhanced medical imaging system. *Acad. Radiol.* **2009**, *16*, 911–917. [CrossRef] [PubMed]
16. Epple, F.M.; Ehn, S.; Thibault, P.; Koehler, T.; Potdevin, G.; Herzen, J.; Pennicard, D.; Graafsma, H.; Noel, P.B.; Pfeiffer, F. Phase unwrapping in spectral X-ray differential phase-contrast imaging with an energy-resolving photon-counting pixel detector. *IEEE Trans. Med. Imaging* **2015**, *34*, 816–823. [CrossRef] [PubMed]
17. Morgan, K.S.; Paganin, D.M.; Siu, K.K.W. X-ray phase imaging with a paper analyzer. *Appl. Phys. Lett.* **2012**, *100*, 124102–124104. [CrossRef]
18. Berujon, S.; Ziegler, E.; Cerbino, R.; Peverini, L. Two-dimensional X-ray beam phase sensing. *Phys. Rev. Lett.* **2012**, *108*, 158102. [CrossRef] [PubMed]

19. Wang, H.; Kashyap, Y.; Cai, B.; Sawhney, K. High energy X-ray phase and dark-field imaging using a random absorption mask. *Sci. Rep.* **2016**, *6*, 30581. [CrossRef] [PubMed]
20. Thuring, T.; Abis, M.; Wang, Z.; David, C.; Stampanoni, M. X-ray phase-contrast imaging at 100 keV on a conventional source. *Sci. Rep.* **2014**, *4*, 5198. [CrossRef] [PubMed]
21. Hernandez, A.M.; Boone, J.M. Tungsten anode spectral model using interpolating cubic splines: Unfiltered X-ray spectra from 20 kV to 640 kV. *Med. Phys.* **2014**, *41*, 042101. [CrossRef] [PubMed]
22. FEPA-Standard. Available online: https://www.fepa-abrasives.com/abrasive-products/grains (accessed on 10 May 2018).
23. The Commonly Used Grading System for Steel Wool Fibre Thickness Can Be Found for Example in the Engineering Toolbox as a Reference. Available online: https://www.engineeringtoolbox.com/steel-wool-grades-d_1619.html (accessed on 10 May 2018).
24. Pan, B.; Xie, H.-M.; Xu, B.-Q.; Dai, F.-L. Performance of sub-pixel registration algorithms in digital image correlation. *Meas. Sci. Technol.* **2006**, *17*, 1615.
25. Zhou, T.; Zdora, M.C.; Zanette, I.; Romell, J.; Hertz, H.M.; Burvall, A. Noise analysis of speckle-based X-ray phase-contrast imaging. *Opt. Lett.* **2016**, *41*, 5490–5493. [CrossRef] [PubMed]
26. Berujon, S.; Wang, H.; Sawhney, K. X-ray multimodal imaging using a random-phase object. *Phys. Rev. A* **2012**, *86*, 063813. [CrossRef]
27. Pan, B.; Xie, H.M.; Wang, Z.Y.; Qian, K.M.; Wang, Z.Y. Study on subset size selection in digital image correlation for speckle patterns. *Opt. Express* **2008**, *16*, 7037–7048. [CrossRef] [PubMed]
28. Scrivener, K.L.; Gartner, E.M. Microstructural gradients in cement paste around aggregate particles. *MRS Proc.* **1987**, *114*. [CrossRef]
29. Leemann, A.; Münch, B.; Gasser, P.; Holzer, L. Influence of compaction on the interfacial transition zone and the permeability of concrete. *Cem. Concr. Res.* **2006**, *36*, 1425–1433. [CrossRef]
30. Bentz, D.P.; Martys, N.S.; Stutzman, P.; Levenson, M.S.; Garboczi, E.J.; Dunsmuir, J.; Schwartz, L.M. X-ray microtomography of an ASTM C109 mortar exposed to sulfate attack. *MRS Proc.* **1995**, *370*, 77–82. [CrossRef]
31. Chotard, T.J.; Boncoeur-Martel, M.P.; Smith, A.; Dupuy, J.P.; Gault, C. Application of X-ray computed tomography to characterise the early hydration of calcium aluminate cement. *Cem. Concr. Compos.* **2003**, *25*, 145–152. [CrossRef]
32. Sarapata, A.; Ruiz-Yaniz, M.; Zanette, I.; Rack, A.; Pfeiffer, F.; Herzen, J. Multi-contrast 3D X-ray imaging of porous and composite materials. *Appl. Phys. Lett.* **2015**, *106*, 154102. [CrossRef]
33. Prade, F.; Fischer, K.; Heinz, D.; Meyer, P.; Mohr, J.; Pfeiffer, F. Time resolved X-ray dark-field tomography revealing water transport in a fresh cement sample. *Sci. Rep.* **2016**, *6*, 29108. [CrossRef] [PubMed]
34. Trtik, P.; Soos, M.; Munch, B.; Lamprou, A.; Mokso, R.; Stampanoni, M. Quantification of a single aggregate inner porosity and pore accessibility using hard X-ray phase-contrast nanotomography. *Langmuir* **2011**, *27*, 12788–12791. [CrossRef] [PubMed]
35. Cuesta, A.; De la Torre, A.; Santacruz, I.; Trtik, P.; Da Silva, J.; Diaz, A.; Holler, M.; Aranda, M. In situ hydration imaging study of a ye'elimite paste by ptychographic X-ray computed tomography. In Proceedings of the 39 International Conference on Cement Microscopy, Toronto, ON, Canada, 10–13 April 2017.
36. Feldkamp, L.A.; Davis, L.C.; Kress, J.W. Practical cone-beam algorithm. *J. Opt. Soc. Am. A* **1984**, *1*, 612–619. [CrossRef]
37. Vlassenbroeck, J.; Dierick, M.; Masschaele, B.; Cnudde, V.; Hoorebeke, L.; Jacobs, P. Software tools for quantification of X-ray microtomography. *Nucl. Instrum. Methods Phys. Res. A* **2007**, *580*, 442–445. [CrossRef]
38. Wang, F.X.; Wang, Y.D.; Wei, G.X.; Du, G.H.; Xue, Y.L.; Hu, T.; Li, K.; Deng, B.; Xie, H.L.; Xiao, T.Q. Speckle-tracking X-ray phase-contrast imaging for samples with obvious edge-enhancement effect. *Appl. Phys. Lett.* **2017**, *111*, 174101. [CrossRef]
39. Wang, H.; Kashyap, Y.; Sawhney, K. From synchrotron radiation to lab source: Advanced speckle-based X-ray imaging using abrasive paper. *Sci. Rep.* **2016**, *6*, 20476. [CrossRef] [PubMed]
40. Zdora, M.-C.; Thibault, P.; Zhou, T.; Koch, F.J.; Romell, J.; Sala, S.; Last, A.; Rau, C.; Zanette, I. X-ray phase-contrast imaging and metrology through unified modulated pattern analysis. *Phys. Rev. Lett.* **2017**, *118*, 203903. [CrossRef] [PubMed]

Journal of
Imaging

MDPI

Article

Single-Shot X-ray Phase Retrieval through Hierarchical Data Analysis and a Multi-Aperture Analyser

Marco Endrizzi *, Fabio A. Vittoria and Alessandro Olivo

Department of Medical Physics and Biomedical Engineering, University College London, Gower Street, London WC1E 6BT, UK; fabio.vittoria.12@ucl.ac.uk (F.A.V.); a.olivo@ucl.ac.uk (A.O.)
* Correspondence: m.endrizzi@ucl.ac.uk

Received: 11 May 2018; Accepted: 5 June 2018; Published: 6 June 2018

Abstract: A multi-aperture analyser set-up was recently developed for X-ray phase contrast imaging and tomography, simultaneously attaining a high sensitivity and wide dynamic range. We present a single-shot image retrieval algorithm in which differential phase and dark-field images are extracted from a single intensity projection. Scanning of the object is required to build a two-dimensional image, because only one pre-sample aperture is used in the experiment reported here. A pure-phase object approximation and a hierarchical approach to the data analysis are used in order to overcome numerical instabilities. The single-shot capability reduces the exposure times by a factor of five with respect to the standard implementation and significantly simplifies the acquisition procedure by only requiring sample scanning during data collection.

Keywords: X-ray imaging; phase-contrast imaging; dark-field imaging

1. Introduction

Non-destructive inspection is often carried out using X-ray radiation because high resolution images can be obtained through significant thicknesses over a wide range of materials. X-ray imaging is applied across a huge variety of fields, such as medicine, materials engineering, biology, and security. Since their introduction more than a hundred years ago, X-ray systems have relied on attenuation to generate contrast and produce an image. When the attenuation contrast is too weak to visualise the internal structure of a sample, phase effects can be exploited to modulate the detected intensity and to enhance the visibility of details that lack sufficient absorption contrast [1]. A number of different approaches have been developed for X-ray Phase Contrast Imaging (XPCI), relying both on laboratory sources and synchrotron radiation facilities, including free-space propagation (with single-distance implementations), crystal analysers, grating interferometry, and speckle-based methods [2–18]. Edge illumination (EI) [19] was developed at a synchrotron facility and later translated to laboratory sources [20]. It has been shown to provide high resolution, quantitative phase, and dark field images [21–23]. Low spatial and temporal coherence is extremely well tolerated by the technique, which is also stable against thermal and mechanical stresses [24–27]. The main concept behind the EI approach is that by strongly shaping the X-ray beam in one direction and inserting a sharp absorbing element before detection, small angular deviations in the direction of propagation of the X-rays are translated into intensity modulations at the detector. When a sample is present in the beam, three main effects can be detected. A certain amount of radiation stops in the sample; this is the sample transmission and reduces the total intensity of the beam that reaches the analyser. The beam is shifted in one direction; this is the sample refraction, and it is translated into an increased or a decreased detected intensity, depending on the direction of shifting. The angular dispersion of the X-ray beam is increased; this is also reflected as a change in intensity depending on the relative arrangement of the structuring and the analysing element.

Building on the EI approach, a multi-aperture analyser set-up was recently developed enabling the simultaneous attainment of high sensitivity and a dynamic range [28,29]. When compared to other full-field imaging techniques, having one pre-sample aperture imposes sample scanning to construct a planar, two-dimensional image. This requirement can be relaxed or eliminated by using a pre-sample mask. One advantage is the possibility of operating single-shot, which can be of interest when the dose to sample or total scanning time need to be optimised. A phase retrieval and data processing scheme is presented, which provides differential phase and dark field images from a single detector exposure.

2. Experimental

The multi-aperture analyser setup is composed of a pre-sample slit for beam shaping and a multi-aperture slit for beam analysis before detection. A schematic of the set-up is shown in Figure 1. For a comparison of the multi-aperture setup against the more conventional single-aperture setup, please refer to the Supplemental Material of ref. [28]. Since, in this case, a single aperture (before the sample) was available, a single image line was acquired for each detector exposure, and a two-dimensional image was built by scanning the sample vertically and collecting multiple exposures. Five apertures in the detector mask were used in this experiment.

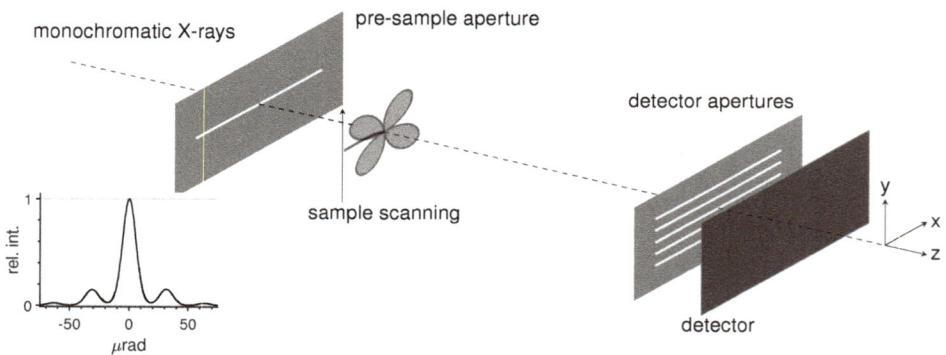

Figure 1. Sketch of the set-up: 20 keV monochromatic synchrotron radiation is amplitude-modulated with a narrow slit, traverses the sample and is analysed by a set of apertures just before the detector. The sample is scanned along the vertical (y) direction and a two-dimensional image is built line-by-line. The illumination function is shown in the inset in the bottom left corner.

The intensity measured in the detector pixels is modulated by scanning the analyser vertically, and is a function of its misalignment $\bar{y}$ with the pre-sample aperture (maximum intensity at $\bar{y} = 0$). This is typically referred to as the illumination function (IF) $L(\bar{y})$ which characterises the imaging system (shown in the inset in the bottom left corner of Figure 1). Once the sample, O, is placed in the beam, the intensity at the detector can be expressed as a convolution between the sample and the IF [28]:

$$I(\bar{y}) \quad = \quad \int L(\bar{y} - y)O(y)dy. \tag{1}$$

It is useful to express I as a sum of Gaussian functions [22] in order to separate the contributions from the (known) IF and the (unknown) sample:

$$I(\bar{y}) = \sum_m \sum_n A_{mn} \exp\left[-\frac{(\bar{y}-\mu_{mn})^2}{2\sigma_{mn}^2}\right], \tag{2}$$

where $L(\bar{y}) = \sum_n (A_n/\sqrt{2\pi\sigma_n^2}) \exp\left[-(\bar{y}-\mu_n)^2/2\sigma_n^2\right]$, $(n = 1\ldots N)$, $O(\bar{y}) = \sum_m (A_m/\sqrt{2\pi\sigma_m^2})$ $\exp\left[-(\bar{y}-\mu_m)^2/2\sigma_m^2\right]$ and $(m = 1\ldots M)$. A, μ and σ are the usual three parameters that specify a Gaussian function, namely its amplitude, mean and width. The parameters are defined according to $\mu_{mn} = \mu_m + \mu_n$, $\sigma_{mn}^2 = \sigma_m^2 + \sigma_n^2$ and $A_{mn} = A_m A_n(1/\sqrt{2\pi\sigma_{mn}^2})$. Equations (1) and (2) hold independently for each image pixel.

The data (used to retrieve the images shown later) were collected at the SYRMEP beamline of the Elettra Synchrotron Facility in Trieste, Italy. The pre-sample aperture, placed at 22 m from the source, was 20 µm wide, whilst the analyser apertures were 23 µm wide with a periodicity of 79 µm and positioned 2.76 m downstream. The sample was positioned at 26 cm from the pre-sample aperture and was scanned in a step-and-shoot fashion. Synchrotron radiation at 20 keV was selected from a bending magnet source by using a double-bounce Si(111) monochromator. The detector was a Photonic Science Ltd. charge-coupled device with a 12.5 µm pixel pitch. Exposures were 500 ms per sample position. The sample was a *Lunaria annua* flower of which two petals were exposed. The field of view was limited horizontally by the width of the pre-sample aperture to 40 mm, while vertically, it was determined by the length of the acquisition scan.

3. Algorithm and Results

The standard acquisition scheme [28] entailed five exposures per sample position, each with the analyser aligned to a different $\bar{y}_i$, therefore selecting different illumination levels. In this work, five intensities wee recorded through each aperture j ($j = 1\ldots5$) for each pixel i without moving the analyser. The intensities recorded with $(\bar{I}_{i,j})$ and without $(L_{i,j})$ the sample in the beam were then compared to extract the sample function, O, at each position in the plane (x,y). This function contains the properties of the sample in terms of transmission (amplitude), refraction (mean) and scattering (width). Retrieval of the function, O, is achieved through a non-linear fitting procedure that minimises $\min_b \sum_j [I(b, \bar{y}_0) - \bar{I}]^2$, where I is the model function of Equation (2), b its set of parameters (specifying the Gaussian model of Equation (2)) and $\bar{I}_j$ is the X-ray intensity measured through each aperture at a single analyser position $\bar{y}_0$. A maximum number of 100 iterations was performed in the search for the optimal parameters. For the single-shot case, only one $\bar{y}$ position of the analyser was used. Reducing the amount of data by a factor of five, compared to the five-shots method previously employed, resulted in numerical instability of the retrieval—the noise in the raw intensity data propagated through the retrieval and compromised the quality of the sample images.

In order to overcome this problem we first imposed a pure phase object constraint. This was justified by observing that at relatively high energy (20 keV) and for a thin petal ($\lesssim 100$ µm) the absorption was estimated to be less than 1%.

A good initial guess is also critical for successful retrieval in the presence of noise. In other words, the optimal values should be obtained with few iterations before the propagation of noise takes over the signal. This problem was addressed through a hierarchical data processing approach. The algorithm worked in the following way: the raw intensity images were heavily binned, providing an image with very few pixels but excellent photon statistics. Phase retrieval was applied to this data, starting the iteration from a flat, inaccurate guess centred around zero. Because of the excellent statistics, the actual noise was extremely small and the retrieval performed well, with fast and robust convergence. The values of the sample function, O, obtained in this way were accurate but had low resolution. They did, however, provide a good starting point for the next iteration. The following iteration step was implemented by taking the retrieved O function, expanding it to match the new binning factor of the raw intensity image and using it as the initial guess for the phase retrieval.

A schematic representation of the algorithm is also reported in Figure 2. The results obtained with this procedure are shown in Figure 3 where differential phase and dark-field images were obtained with three different binning factors: 50×50, 20×20 and 4×4. As a rule of thumb, we roughly halved the binning factor at each iteration.

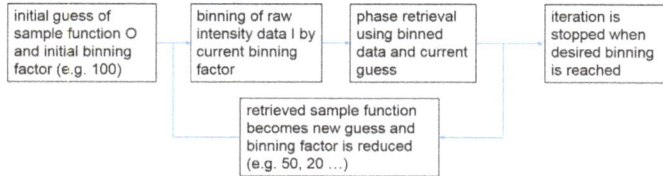

initial guess of sample function O and initial binning factor (e.g. 100)	binning of raw intensity data I by current binning factor	phase retrieval using binned data and current guess	iteration is stopped when desired binning is reached

retrieved sample function becomes new guess and binning factor is reduced (e.g. 50, 20 ...)

Figure 2. Schematic representation of the algorithm.

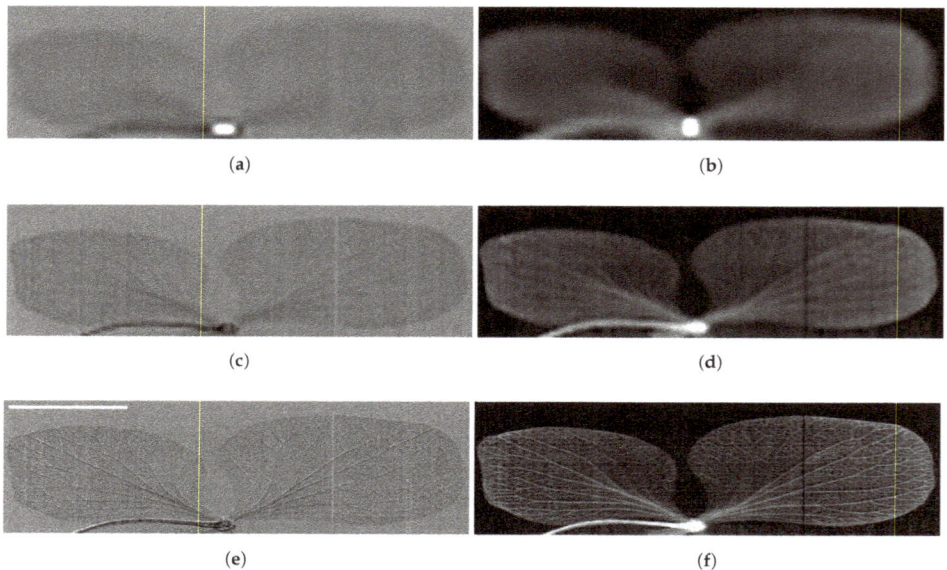

Figure 3. Single-shot X-ray (**a**,**c**,**e**) differential phase and (**b**,**d**,**f**) dark-field images of a flower obtained through hierarchical data analysis. To overcome numerical instabilities, the raw intensity images were processed by iteratively decreasing binning: (**a**,**b**) used 50×50, (**c**,**d**) used 20×20 and (**e**,**f**) used 4×4. The scale bar in panel (**e**) is 1 cm.

This procedure stabilised the retrieval and enabled single-shot, high resolution images to be extracted from the data. Some stripe artefacts were still visible in the images; these are most likely due to the type of acquisition that, in this configuration, requires a vertical scan. An ill-behaved pixel affects a whole image column, and its effect therefore becomes much more visible with respect to a standard two-dimensional pixel arrangement.

4. Conclusions

In summary, we have presented a hierarchical data processing algorithm that enables single-shot hard X-ray phase and dark-field imaging retrieval. The experimental set-up is based on the multi-aperture analyser that, when used in this configuration, does not require any instrumentation movement during data acquisition. In the configuration used here, the sample has to be scanned through the laminar beam created by the pre-sample aperture in order to build a two-dimensional image and this is due to using only a single pre-sample aperture. This data acquisition and analysis scheme could be of interest when exposure times have to be reduced in favour of dose or total scanning time, or when having stationary optical elements offers improved stability and accuracy in the measurement.

Author Contributions: M.E., F.A.V. and A.O. conceived, designed and performed the experiments; M.E. analysed the data and wrote the paper with contributions from F.A.V. and A.O.

Acknowledgments: M.E. was supported by the Royal Academy of Engineering under the RAEng Research Fellowships scheme. F.A.V. was supported by the Royal Academy of Engineering and the Office of the Chief Science Adviser for National Security under the UK Intelligence Community Postdoctoral Fellowship Programme. This work was also supported by the EPSRC (Grant EP/M507970/1). We thank Elettra Sincrotrone Trieste for access to SYRMEP beamline (proposal 20140147) that contributed to the results presented here.

Conflicts of Interest: The authors declare no conflict of interest. The founding sponsors had no role in the design of the study; in the collection, analyses, or interpretation of data; in the writing of the manuscript, and in the decision to publish the results.

Abbreviations

The following abbreviations are used in this manuscript:

XPCI   X-ray Phase Contrast Imaging
IF   Illumination Function

References

1. Endrizzi, M. X-ray phase-contrast imaging. *Nucl. Instrum. Methods Phys. Res. Sect. A Accel. Spectrom. Detect. Assoc. Equip.* **2018**, *878*, 88–98.
2. Bonse, U.; Hart, M. An X-ray interferometer. *Appl. Phys. Lett.* **1965**, *6*, 155–156.
3. Goetz, K.; Foerster, E.; Zaumseil, P.; Kalashnikov, M.P.; Mikhailov, I.A.; Sklizkov, G.V.; Fedotov, S.I. Measurements of the parameters of shell targets for laser thermonuclear fusion using an X-ray schlieren method. *Kvantovaia Elektron. Mosc.* **1979**, *6*, 1037–1042.
4. Davis, T.J.; Gao, D.; Gureyev, T.E.; Stevenson, A.W.; Wilkins, S.W. Phase-contrast imaging of weakly absorbing materials using hard X-rays. *Nature* **1995**, *373*, 595–598.
5. Ingal, V.N.; Beliaevskaya, E.A. X-ray plane-wave topography observation of the phase contrast from a non-crystalline object. *J. Phys. D Appl. Phys.* **1995**, *28*, 2314–2317.
6. Wilkins, S.W.; Gureyev, T.E.; Gao, D.; Pogany, A.; Stevenson, A.W. Phase-contrast imaging using polychromatic hard X-rays. *Nature* **1996**, *384*, 335–338.
7. Chapman, D.; Thomlinson, W.; Johnston, R.E.; Washburn, D.; Pisano, E.; Gmür, N.; Zhong, Z.; Menk, R.; Arfelli, F.; Sayers, D. Diffraction enhanced X-ray imaging. *Phys. Med. Biol.* **1997**, *42*, 2015–2025.
8. Clauser, J.F. Ultrahigh resolution interferometric X-ray imaging. U.S. Patent 5,812,629, 22 September 1998.
9. David, C.; Nohammer, B.; Solak, H.H.; Ziegler, E. Differential X-ray phase contrast imaging using a shearing interferometer. *Appl. Phys. Lett.* **2002**, *81*, 3287–3289.
10. Paganin, D.; Mayo, S.C.; Gureyev, T.E.; Miller, P.R.; Wilkins, S.W. Simultaneous phase and amplitude extraction from a single defocused image of a homogeneous object. *J. Microsc.* **2002**, *206*, 33–40.
11. Momose, A.; Kawamoto, S.; Koyama, I.; Hamaishi, Y.; Takai, K.; Suzuki, Y. Demonstration of X-ray Talbot Interferometry. *Jpn. J. Appl. Phys.* **2003**, *42*, L866.
12. Mayo, S.C.; Sexton, B. Refractive microlens array for wave-front analysis in the medium to hard X-ray range. *Opt. Lett.* **2004**, *29*, 866–868.
13. Pfeiffer, F.; Weitkamp, T.; Bunk, O.; David, C. Phase retrieval and differential phase-contrast imaging with low-brilliance X-ray sources. *Nat. Phys.* **2006**, *2*, 258–261.
14. De Jonge, M.D.; Hornberger, B.; Holzner, C.; Legnini, D.; Paterson, D.; McNulty, I.; Jacobsen, C.; Vogt, S. Quantitative Phase Imaging with a Scanning Transmission X-ray Microscope. *Phys. Rev. Lett.* **2008**, *100*, 163902.

15. Wen, H.; Bennett, E.E.; Hegedus, M.M.; Rapacchi, S. Fourier X-ray Scattering Radiography Yields Bone Structural Information1. *Radiology* **2009**, *251*, 910–918.
16. Morgan, K.S.; Paganin, D.M.; Siu, K.K.W. X-ray phase imaging with a paper analyzer. *Appl. Phys. Lett.* **2012**, *100*, 124102.
17. Wang, H.; Kashyap, Y.; Sawhney, K. Hard-X-ray directional dark-field imaging using the speckle scanning technique. *Phys. Rev. Lett.* **2015**, *114*, 103901.
18. Miao, H.; Panna, A.; Gomella, A.A.; Bennett, E.E.; Znati, S.; Chen, L.; Wen, H. A universal moiré effect and application in X-ray phase-contrast imaging. *Nat. Phys.* **2016**, *12*, 830–834.
19. Olivo, A.; Arfelli, F.; Cantatore, G.; Longo, R.; Menk, R.H.; Pani, S.; Prest, M.; Poropat, P.; Rigon, L.; Tromba, G.; et al. An innovative digital imaging set-up allowing a low-dose approach to phase contrast applications in the medical field. *Med. Phys.* **2001**, *28*, 1610–1619.
20. Olivo, A.; Speller, R. A coded-aperture technique allowing X-ray phase contrast imaging with conventional sources. *Appl. Phys. Lett.* **2007**, *91*, 074106.
21. Munro, P.R.; Ignatyev, K.; Speller, R.D.; Olivo, A. Phase and absorption retrieval using incoherent X-ray sources. *Proc. Natl. Acad. Sci. USA* **2012**, *109*, 13922–13927.
22. Endrizzi, M.; Diemoz, P.C.; Millard, T.P.; Jones, J.L.; Speller, R.D.; Robinson, I.K.; Olivo, A. Hard X-ray dark-field imaging with incoherent sample illumination. *Appl. Phys. Lett.* **2014**, *104*, 024106.
23. Endrizzi, M.; Vittoria, F.A.; Diemoz, P.C.; Lorenzo, R.; Speller, R.D.; Wagner, U.H.; Rau, C.; Robinson, I.K.; Olivo, A. Phase-contrast microscopy at high X-ray energy with a laboratory setup. *Opt. Lett.* **2014**, *39*, 3332–3335.
24. Munro, P.R.T.; Ignatyev, K.; Speller, R.D.; Olivo, A. Source size and temporal coherence requirements of coded aperture type X-ray phase contrast imaging systems. *Opt. Express* **2010**, *18*, 19681.
25. Endrizzi, M.; Vittoria, F.A.; Kallon, G.; Basta, D.; Diemoz, P.C.; Vincenzi, A.; Delogu, P.; Bellazzini, R.; Olivo, A. Achromatic approach to phase-based multi-modal imaging with conventional X-ray sources. *Opt. Express* **2015**, *23*, 16473–16480.
26. Millard, T.P.; Endrizzi, M.; Ignatyev, K.; Hagen, C.K.; Munro, P.R.T.; Speller, R.D.; Olivo, A. Method for automatization of the alignment of a laboratory based X-ray phase contrast edge illumination system. *Rev. Sci. Instrum.* **2013**, *84*, 083702.
27. Endrizzi, M.; Basta, D.; Olivo, A. Laboratory-based X-ray phase-contrast imaging with misaligned optical elements. *Appl. Phys. Lett.* **2015**, *107*, 124103.
28. Endrizzi, M.; Vittoria, F.A.; Rigon, L.; Dreossi, D.; Iacoviello, F.; Shearing, P.R.; Olivo, A. X-ray Phase-Contrast Radiography and Tomography with a Multiaperture Analyzer. *Phys. Rev. Lett.* **2017**, *118*, 243902.
29. Endrizzi, M.; Vittoria, F.; Brombal, L.; Longo, R.; Zanconati, F.; Olivo, A. X-ray phase-contrast tomography of breast tissue specimen with a multi-aperture analyser synchrotron set-up. *J. Instrum.* **2018**, *13*, C02004.

Journal of
Imaging

MDPI

Article

Optimization Based Evaluation of Grating Interferometric Phase Stepping Series and Analysis of Mechanical Setup Instabilities

Jonas Dittmann [1,*], Andreas Balles [1] and Simon Zabler [1,2]

[1] Lehrstuhl für Röntgenmikroskopie, Universität Würzburg, Josef-Martin-Weg 63, 97074 Würzburg, Germany; andreas.balles@physik.uni-wuerzburg.de (A.B.); simon.zabler@iis.fraunhofer.de (S.Z.)
[2] Fraunhofer EZRT, NCTS Group Würzburg, Josef-Martin-Weg 63, 97074 Würzburg, Germany
* Correspondence: jonas.dittmann@physik.uni-wuerzburg.de

Received: 30 April 2018; Accepted: 4 June 2018; Published: 7 June 2018

Abstract: The diffraction contrast modalities accessible by X-ray grating interferometers are not imaged directly but have to be inferred from sine-like signal variations occurring in a series of images acquired at varying relative positions of the interferometer's gratings. The absolute spatial translations involved in the acquisition of these phase stepping series usually lie in the range of only a few hundred nanometers, wherefore positioning errors as small as 10 nm will already translate into signal uncertainties of 1–10 % in the final images if not accounted for. Classically, the relative grating positions in the phase stepping series are considered input parameters to the analysis and are, for the Fast Fourier Transform that is typically employed, required to be equidistantly distributed over multiples of the gratings' period. In the following, a fast converging optimization scheme is presented simultaneously determining the phase stepping curves' parameters as well as the actually performed motions of the stepped grating, including also erroneous rotational motions which are commonly neglected. While the correction of solely the translational errors along the stepping direction is found to be sufficient with regard to the reduction of image artifacts, the possibility to also detect minute rotations about all axes proves to be a valuable tool for system calibration and monitoring. The simplicity of the provided algorithm, in particular when only considering translational errors, makes it well suitable as a standard evaluation procedure also for large image series.

Keywords: X-ray imaging; Talbot–Lau interferometer; grating interferometer; phase contrast imaging; darkfield imaging; phase stepping; optimization

1. Introduction

X-ray grating interferometry [1,2] facilitates access to new contrast modalities in laboratory X-ray imaging setups and has by now been implemented by many research groups after the seminal publication by Pfeiffer et al. in 2006 [3]. The additional information on X-ray refraction ("differential phase contrast") and ultra small angle scattering ("darkfield contrast") properties of a sample that can be obtained promises both increased sensitivity to subtle material variations as well as insights into the samples' substructure below the spatial resolution of the acquired images.

In contrast to classic X-ray imaging, the absorption, differential phase and darkfield contrasts are not imaged directly but are encoded in sinusoidal intensity variations arising at each detector pixel when shifting the interferometer's gratings relative to each other perpendicular to the beam path and the grating bars. A crucial step in the generation of respective absorption, phase and darkfield images therefore is the analysis of the commonly acquired phase stepping series, which is the subject of the present article. Respective examples are shown in Figures 1 and 2.

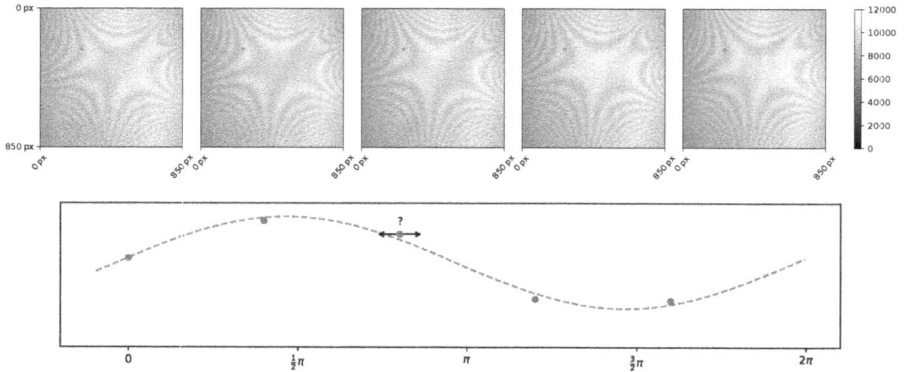

Figure 1. Example for a grating interferometric phase stepping series (without sample, cf. Section 2.3 for experimental details). The intensity variation throughout the series is visually most perceivable in the center. The Moiré fringes are caused by imperfectly matched gratings and will translate to reference offset phases of the sinusoid curves found at each detector pixel (cf. Figure 2). The bottom panel shows a corresponding phase stepping curve for the pixel marked orange in the above image series. The sampling positions are subject to an unknown error.

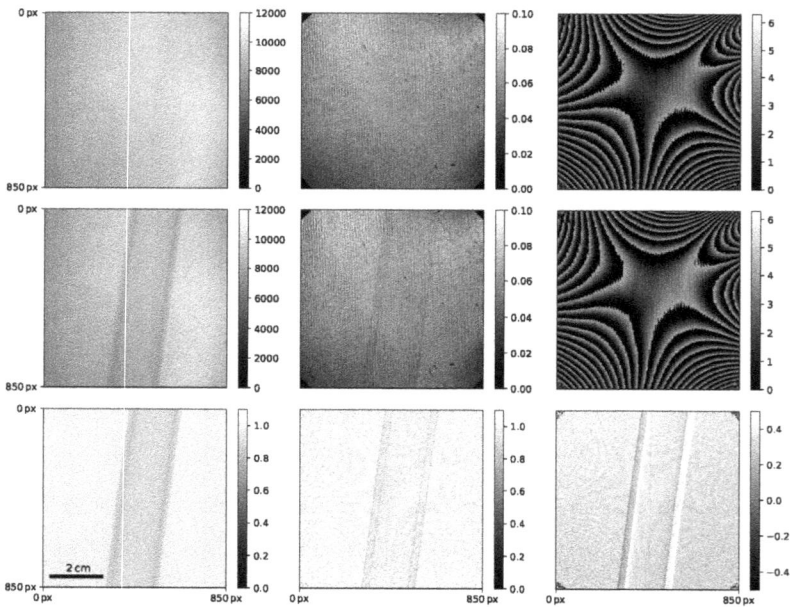

Figure 2. From left to right: Transmission (sinusoid mean), visibility (ratio of sinusoid amplitude and mean) and phase images derived from phase stepping series, as shown in Figure 1. The first two rows show acquisitions with and without sample (a piece of plastic hose of 2 cm diameter), while the last row shows the sample images (center row) normalized with respect to the empty beam images (first row). Positioning errors in the phase stepping procedure cause the Moiré pattern of the reference phase image to translate into the final results.

In principle, the images within such phase stepping series are sampled at about 5–10 different relative grating positions equidistantly distributed over multiples of the gratings' period such that the expected sinusoids for each detector pixel may be characterized by standard Fourier decomposition [1–3]. The zeroth order term represents the mean transmitted intensity (as in classic X-ray imaging), while the first order terms encode phase shift and amplitude of the sinusoid. The ratio of amplitude and mean (generally referred to as "visibility") is here related to scattering and provides the darkfield contrast. Higher order terms correspond to deviations from the sinusoid model mainly due to the actual grating profiles and are usually not considered.

Given typical grating periods in the range of 2–10 micrometers, the actually performed spatial translations lie in the range of 200–2000 nanometers. Particular for the smaller gratings, positioning errors as small as 10–20 nm imply relative phase errors in the range of 5–10 percent, causing uncertainties in the derived quantities in the same order of magnitude. The propagation of noise within the sampling positions onto the extracted signals has, e.g., been studied by Revol et al. [4], and first results for the determination of the actual sampling positions from the available image series were shown by Seifert et al. [5] using methods by Vargas et al. [6] from the context of visible light interferometry based on a decomposition of the images into three basis images (to be found by means of principal component analysis and subsequent optimization) corresponding to the transmission image and the amplitude modulated sine and cosine of the phase image. Kaeppler et al. [7] optimized the phase stepping positions by means of minimizing an objective function penalizing variations within the visibility and differential phase images that arise from phase stepping errors. For the case of simultaneous grating stepping and object rotation, von Teuffenbach et al. [8] recently reported the inference of stepping errors as part of a maximum likelihood tomographic reconstruction procedure proposed by Ritter et al. [9]. Finally, the authors just became aware of the parallel work by De Marco et al. [10] who investigated the origination of the particular image artifacts caused by phase stepping jitter in order to derive a postprocessing artifact reduction technique.

Especially with regard to the quantitative analysis of darkfield signals [11–13], the reduction or elimination of stepping jitter induced errors is highly desirable. This also applies to the directional darkfield [14,15] and anisotropic darkfield tomography techniques [16–21], which rely on the comparability of multiple darkfield images to relate signal variations to the anisotropy and orientation of scatterers. As the tomographic methods furthermore require large amounts (10^3–10^4) of projections, phase stepping series processing efficiency is also an issue. With regard to system calibration and monitoring, the inference of the actual stepping motions from the phase stepping series is of interest.

The present article proposes a simple iterative optimization algorithm both for the fitting of irregularly sampled sinusoids and in particular for the determination of the actual sampling positions. The use of only basic mathematical operations eases straightforward implementations on arbitrary platforms. Besides uncertainties in the lateral stepping motion, the remaining mechanical degrees of freedom (magnification/expansion and rotations) are also considered. The proposed techniques will be demonstrated on a typical data set.

2. Methods

The task of sinusoid fitting with imprecisely known sampling locations will be partitioned into two separate optimization problems considering either only the sinusoid parameters or only the sampling positions while temporarily fixing the respective other set of parameters. Alternating both optimization tasks will minimize the joint objective function in few iterations. The technique is further extended to an objective function considering spatially varying sampling positions to also model (erroneous) relative grating rotations and magnifications.

2.1. Sinusoid Fitting

The sinusoid is defined as a function $\tilde{y}(\phi)$ of phase ϕ parameterized by a constant offset o, an amplitude a and a phase offset ϕ_0:

$$\tilde{y}(\phi) = o + a\sin(\phi - \phi_0). \tag{1}$$

For the purpose of fitting the sinusoid model to given data samples, the equivalent representation as a linear combination of trigonometric basis functions (i.e., as a first order Fourier series) will be more convenient:

$$
\begin{aligned}
o + a\sin(\phi - \phi_0) &= o + a\left(\cos\phi_0\sin\phi - \sin\phi_0\cos\phi\right) \\
&= o + a_s\sin\phi + a_c\cos\phi \\
&= (o, a_s, a_c) \cdot (1, \sin\phi, \cos\phi)^T \quad ,
\end{aligned}
\tag{2}
$$

parameterized by the constant offset o and two auxiliary amplitudes a_s and a_c with the following identities:

$$
\begin{aligned}
a &= \sqrt{a_s^2 + a_c^2} \\
\sin\phi_0 &= -a_c/a \\
\cos\phi_0 &= a_s/a \\
\phi_0 &= \arctan2(-a_c, a_s).
\end{aligned}
\tag{3}
$$

The parameters o, a_c, and a_s of the recast sinusoid model corresponding to the least squares fit to given data samples (ϕ_i, y_i) enumerated by i may be determined by means of the following iterative scheme (introducing the superscript iteration index k):

$$
\begin{aligned}
o^{(0)}, a_c^{(0)}, a_s^{(0)} &= 0, 0, 0 \\
\tilde{y}_i^{(k)} &= o^{(k)} + a_c^{(k)}\cos\phi_i + a_s^{(k)}\sin\phi_i \\
o^{(k+1)} &= o^{(k)} + \frac{1}{N}\sum_i(y_i - \tilde{y}_i^{(k)}) \\
a_s^{(k+1)} &= a_s^{(k)} + \frac{2}{N}\sum_i(y_i - \tilde{y}_i^{(k)})\sin\phi_i \\
a_c^{(k+1)} &= a_c^{(k)} + \frac{2}{N}\sum_i(y_i - \tilde{y}_i^{(k)})\cos\phi_i \quad ,
\end{aligned}
\tag{4}
$$

with the factors of $1/N$ and $2/N$ accounting for the normalization of the respective basis functions and N being the amount of samples (ϕ_i, y_i) enumerated by i. The intermediate variables $\tilde{y}_i^{(k)}$ denote the respective ordinate values of the sinusoid model for iteration k and abcissas ϕ_i. The differences $y_i - \tilde{y}_i^{(k)}$ thus correspond to the residuals at iteration k. The scheme reduces to classic Fourier analysis for the case of the abscissas ϕ_i being equidistantly distributed over multiples of 2π and converges within the first iteration in that case. As the incremental updates to $o^{(k)}$, $a_s^{(k)}$ and $a_c^{(k)}$ are proportional to the respective derivatives of the ℓ_2 error $\sum_i(o^{(k)} + a_c^{(k)}\cos\phi_i + a_s^{(k)}\sin\phi_i - y_i)^2$, the fixpoint of the iteration will be the least squares fit also in all other cases.

For a stopping criterion, the relative error reduction

$$\Delta_{\ell_2} = \frac{\sqrt{\sum_i\left(y_i - \tilde{y}_i^{(k-1)}\right)^2} - \sqrt{\sum_i\left(y_i - \tilde{y}_i^{(k)}\right)^2}}{\sqrt{\sum_i\left(y_i - \tilde{y}_i^{(k)}\right)^2}} \tag{5}$$

may be tracked. It is typically found to fall below 0.1% within 10–20 iterations given only slightly noisy data (noise sigma three orders of magnitude smaller than sinusoid amplitude) and within less than 10 iterations for most practical cases. For the special case of equidistributed ϕ_i on multiples of 2π, it will immediately drop to 0 after the first iteration. In practice, a fixed amount of iterations in the range of five to fifteen will therefore be adequate as stopping criterion as well.

2.2. Phase Step Optimization

An underlying assumption of the previously described least squares fitting procedure is the certainty of the abscissas, i.e., the set of phases ϕ_i at which the ordinates y_i have been sampled. As the sampling positions are themselves subject to experimental uncertainties (arising from the mechanical precision of the involved actuators), a further optimization step will be introduced that minimizes the least squares error of the sinusoid fit over deviations $\Delta\phi_i$ from the intended sampling positions ϕ_i. These deviations are expected to be smaller than the typical stepping increment $\phi_{i+1} - \phi_i$. While this procedure obviously results in overfitting when considering only a single phase stepping curve (PSC), it becomes a well-defined error minimization problem when regarding large sets of PSCs sharing the same actual sampling positions $\phi_i + \Delta\phi_i$. In other words, an approach to the minimization problem

$$o_j, a_j, \phi_{0,j}, \Delta\phi_i = \operatorname*{argmin}_{o_j, a_j, \phi_{0,j}, \Delta\phi_i} \sum_{i,j} \left(o_j + a_j \sin(\phi_i + \Delta\phi_i - \phi_{0,j}) - y_{ji} \right)^2 \tag{6}$$

shall be considered, with j indexing phase stepping curves captured by different detector pixels at identical stepping positions $\phi_i + \Delta\phi_i$.

To derive an optimization procedure for the sampling positions, first the fictive case of a perfectly sinusoid PSC with negligible statistical error on the ordinate (the sampled values) shall be considered. Ignoring for now the fact that least squares fits commonly assume only the ordinates to be affected by noise, a least squares fit shall be used to preliminarily determine the parameters of the sinusoid described by the observed data. Assuming then that inconsistencies of the observed data with the model are due to errors on the sampling locations, deviations from their intended positions are given by the data points' lateral distances from the sinusoid curve (cf. Figure 3). Finally, the actual systematic deviations of the sampling locations can be found by averaging over the respective results for a large set of PSCs sampled simultaneously. This information can be fed back into the original sinusoid fit, which then again allows the refinement of the current estimate of the true sampling positions, finally resulting in an iterative procedure alternatingly optimizing the sinusoid parameters and the actual sampling locations (cf. Algorithm 1).

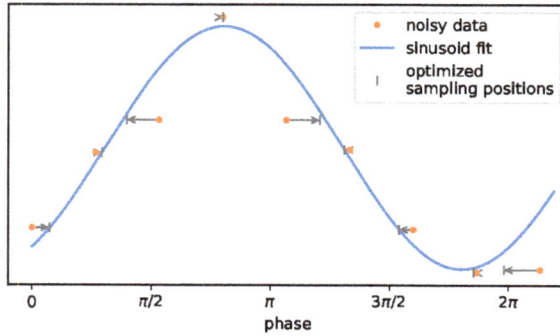

Figure 3. Optimization of the actual (in contrast to the intended) sampling positions by Equation (14) with respect to a previous sinusoid fit based on the temporary assumption that deviations from the sinusoid model are mainly due to errors on the sampling positions rather than the sampled ordinate values. Averaging of respective phase deviations found for large sets of measurements will finally allow the differentiation of systematic deviations from statistical noise.

Algorithm 1 Least squares optimization of shared abscissa values ϕ_i for simultaneous sinusoid fits to ordinate samples y_{ji} belonging to independent curves j sampled at identical positions ϕ_i. This represents a special case of Algorithm 2 with spatially invariant sampling phases. The relaxation parameter $\lambda_k \in (0;1]$ may be chosen <1 if damping of the updates to $\phi_i^{(k)}$ is desired. For the intermediate argmin operations, see Equations (2)–(5).

1:   ϕ_i, y_{ji}: input data

2:   $m_0 \leftarrow \frac{1}{2}$ $\triangleright$ upper limit to $\Delta\phi_{ji}, m_0 \in (0;1.38]$

3:   $\phi_i^{(0)} \leftarrow \phi_i$

4:   $o_j^{(0)}, a_j^{(0)}, \phi_{0,j}^{(0)} \leftarrow \underset{o,a,\phi_0}{\mathrm{argmin}} \sum_i \left(o + a\sin(\phi_i^{(0)} - \phi_0) - y_{ji} \right)^2$ $\triangleright$ initialization

5:   **for** $k = 0 .. k_{max}$ **do**

6:      $\Delta\phi_i^{(k)} \leftarrow \dfrac{\sum_j \mathrm{softlimit}\left(\cos(\phi_i^{(k)} - \phi_{0,j}^{(k)})\left(a_j^{(k)}(y_{ji} - o_j^{(k)}) - a_j^2\sin(\phi_i^{(k)} - \phi_{0,j}^{(k)}) \right), (a_j^{(k)})^2 m_0 \cos^4(\phi_i^{(k)} - \phi_{0,j}^{(k)}) \right)}{\sum_j (a_j^{(k)})^2 \cos^2(\phi_i^{(k)} - \phi_{0,j}^{(k)})}$

7:      $\phi_i^{(k+1)} \leftarrow \phi_i^{(k)} + \lambda_k \Delta\phi_i^{(k)}$

8:      $o_j^{(k+1)}, a_j^{(k+1)}, \phi_{0,j}^{(k+1)} \leftarrow \underset{o_j,a_j,\phi_{0,j}}{\mathrm{argmin}} \sum_i \left(o_j + a_j\sin(\phi_i^{(k+1)} - \phi_{0,j}) - y_{ji} \right)^2$

9:   **end for.**

Algorithm 2 Simultaneous least squares optimization of abscissa values ϕ_{ji} and sinusoid fits to ordinate samples y_{ji} belonging to independent curves j sampled at positions $\phi_{ji} = \phi_i(j)$ with $\phi_i(j)$ being a slowly varying polynomial with respect to the spatial coordinates $h(j)$ and $v(j)$ accounting for the expected effects due to translations, magnification and rotations of an interferometer's gratings. The procedure reduces to Algorithm 1 when considering only the zeroth order term of $\phi_i(j)$.

1: ϕ_i, y_{ji}: input data
2: $m_0 \leftarrow \frac{1}{2}$ ▷ upper limit to $\Delta\phi_{ji}$, $m_0 \in (0; 1.38]$
3: $\phi_{ji}^{(0)} \leftarrow \phi_i \forall j$ ▷ initialization of sampling phases with intended values
4: $o_j^{(0)}, a_j^{(0)}, \phi_{0,j}^{(0)} \leftarrow \operatorname*{argmin}_{o,a,\phi_0} \sum_i \left(o + a\sin(\phi_i^{(0)} - \phi_0) - y_{ji}\right)^2$ ▷ initial sinusoid fits
5: **for** $k = 0..k_{max}$ **do**
6: $\Delta\phi_{ji}^{(k)} \leftarrow \begin{cases} 0 & a_j^{(k)}\cos(\phi_{ji}^{(k)} - \phi_{0,j}^{(k)}) = 0 \\ \text{softlimit}\left(\frac{(y_{ji} - o_j^{(k)})/a_j^{(k)} - \sin(\phi_{ji}^{(k)} - \phi_{0,j}^{(k)})}{\cos(\phi_{ji}^{(k)} - \phi_{0,j}^{(k)})}, m_0\cos^2(\phi_{ji}^{(k)} - \phi_{0,j}^{(k)})\right) & \text{else} \end{cases}$
7: $w_{ji}^{(k)} \leftarrow a_j^{(k)\,2}\cos^2(\phi_{ji}^{(k)} - \phi_{0,j}^{(k)})$
8: $\Delta\phi_i^{(k)}, \nabla_h\phi_i^{(k)}, \nabla_v\phi_i^{(k)}, \nabla_{hv}\phi_i^{(k)}, \nabla_{h2}\phi_i^{(k)} \leftarrow \operatorname*{argmin}_{\Delta\phi_i, \nabla_h\phi_i, \nabla_v\phi_i, \nabla_{hv}\phi_i, \nabla_{h2}\phi_i} \sum_j w_{ji}\left(\Delta\phi_i(j) - \Delta\phi_{ji}^{(k)}\right)^2$
          ▷ for $\Delta\phi_i(j)$, cf. Equation (19)
9: $\phi_{ji}^{(k+1)} \leftarrow \phi_{ji}^{(k)} + \Delta\phi_i^{(k)} + \nabla_h\phi_i^{(k)}(h - h_0) + \nabla_v\phi_i^{(k)}(v - v_0) + \nabla_{hv}\phi_i^{(k)}(h - h_0)(v - v_0) + \nabla_{h2}\phi_i^{(k)}(h - v_0)^2$
10: $o_j^{(k+1)}, a_j^{(k+1)}, \phi_{0,j}^{(k+1)} \leftarrow \operatorname*{argmin}_{o_j, a_j, \phi_{0,j}} \sum_i \left(o_j + a_j\sin(\phi_{ji}^{(k+1)} - \phi_{0,j}) - y_{ji}\right)^2$
11: **end for**.

2.2.1. Determination of Individual Phase Deviations

Starting with an initial sinusoid fit

$$o, a, \phi_0 = \operatorname*{argmin}_{o,a,\phi_0} \sum_i (o + a\sin(\phi_i - \phi_0) - y_i)^2 \,, \tag{7}$$

to given data samples (ϕ_i, y_i), the residual sum of squares shall be minimized over deviations $\Delta\phi_i$ to the abscissas ϕ_i while keeping the sinusoid model parameters o, a and ϕ_0 fixed:

$$\Delta\phi_i = \operatorname*{argmin}_{\Delta\phi_i} \sum_i (o + a\sin(\phi_i + \Delta\phi_i - \phi_0) - y_i)^2 \,. \tag{8}$$

For reasons of better readability, the detector pixel index (or equivalently PSC index) j of the sinusoid parameters o, a, ϕ_0 and the data samples y_i has been omitted here and is explicitly added again to the final results at the end of this subsection.

Equation (8) is solved when the derivative of the objective function with respect to $\Delta\phi_i$ vanishes:

$$0 = \frac{\mathrm{d}}{\mathrm{d}\Delta\phi_i} \sum_i (o + a\sin(\phi_i + \Delta\phi_i - \phi_0) - y_i)^2$$
$$0 = 2(o - y_i + a\sin(\phi_i + \Delta\phi_i - \phi_0))\, a\cos(\phi_i + \Delta\phi_i - \phi_0)$$
$$\text{for } a\cos(\phi_i + \Delta\phi_i - \phi_0) \neq 0: \tag{9}$$
$$0 = (o - y_i + a\sin(\phi_i + \Delta\phi_i - \phi_0))$$
$$\text{for } \Delta\phi_i \ll \pi:$$
$$0 \approx (o - y_i + a\sin(\phi_i - \phi_0) + \Delta\phi_i a\cos(\phi_i - \phi_0))$$

where the last step is a first order Taylor expansion with respect to $\Delta\phi_i$. This directly leads to the following expression for $\Delta\phi_i$:

$$\Delta\phi_i \approx \frac{1}{\cos(\phi_i - \phi_0)}\left(\frac{y_i - o}{a} - \sin(\phi_i - \phi_0)\right) \quad \text{for } \Delta\phi_i \ll \pi \text{ and } a\cos(\phi_i - \phi_0) \neq 0 \,, \tag{10}$$

where the earlier condition $a\cos(\phi_i + \Delta\phi_i - \phi_0) \neq 0$ is approximated to be satisfied when $a\cos(\phi_i - \phi_0) \neq 0$. The restriction to cases with $a\cos(\phi_i - \phi_0) \neq 0$ can be intuitively understood when recalling that $\cos(\phi_i - \phi_0) = 0$ implies a maximum or minimum of the sinusoid and $a = 0$ means that it is constant (ϕ_i independent), in both of which cases there is no sensible choice for $\Delta\phi_i \neq 0$. The constraint on the result, $\Delta\phi_i \ll \pi$, can simply be taken into account by means of a limiting function parameterized by a maximal absolute value $m \geq 0$ such as

$$\text{softlimit}(\Delta\phi_i, m \geq 0) = \begin{cases} 0 & m = 0 \\ m \tanh\left(\frac{\Delta\phi_i}{m}\right) & \text{else} \end{cases}, \tag{11}$$

which converges to the identity function for $\Delta\phi_i \ll m$ and is bounded at $\pm m$. The choice of m in this case depends on the validity range of the linear approximation of $\sin(\phi_i + \Delta\phi_i - \phi_0)$ with respect to $\Delta\phi_i$ about $\phi_i - \phi_0$, which obviously depends on the magnitude of the curvature of the sinusoid at this point, as illustrated in Figure 4. This may be accounted for by introducing a suitable $\phi_i - \phi_0$ dependence to m denoted by the respective argument:

$$m(\phi_i - \phi_0) = m_0 \cos^2(\phi_i - \phi_0). \tag{12}$$

$m(\phi_i - \phi_0)$ reaches its largest value m_0 at $\phi_i - \phi_0 = 0$ (where $\sin(\phi_i - \phi_0)$ is actually linear) and smoothly drops to 0 for $\cos(\phi_i - \phi_0) = 0$, in which case both the sine and its curvature are extremal and $\Delta\phi_i$ shall and will be limited to 0. The upper bound for $m(\phi_i - \phi_0)$ and thus for the choice of m_0 is defined by the range of values $\Delta\phi_i \in [-m_0 \cos^2(\phi_i - \phi_0), +m_0 \cos^2(\phi_i - \phi_0)]$ over which $\sin(\Delta\phi_i + \phi_i - \phi_0)$ is actually invertible (cf. Figure 4). The sine function $\sin(\vartheta)$ is locally invertible over intervals of $\vartheta \in [n\pi - \frac{\pi}{2}, n\pi + \frac{\pi}{2}]$ with n being an integer number and ϑ denoting the argument ($\vartheta = \Delta\phi_i + \phi_i - \phi_0$). For the determination of m_0 it is sufficient to consider the case $n = 0$:

$$m_0 \cos^2(\vartheta) \leq \frac{\pi}{2} - |\vartheta| \quad \forall \vartheta \in [-\frac{\pi}{2}, +\frac{\pi}{2}]$$
$$m_0 \lesssim 1.38. \tag{13}$$

For $m_0 = 1.38$, the linear approximation used in Equation (9) deviates by up to 40%. The deviation is limited to 20% or 5% for $m_0 = 1$ and $m_0 = \frac{1}{2}$, respectively.

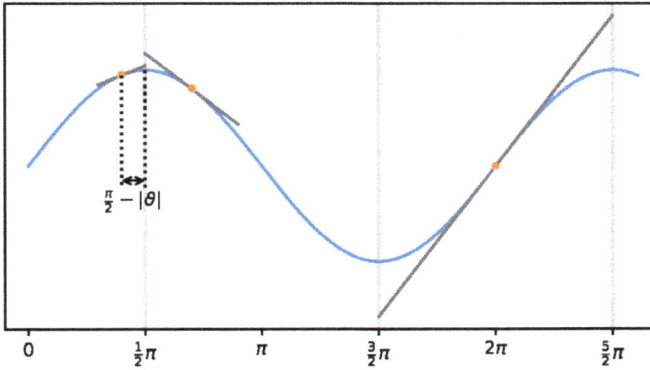

Figure 4. Illustration of the maximal sensible ranges for linear approximations of a sinusoid. Vertical lines at the turning points indicate the boundaries of monotone sections that should never be crossed by linear approximations of the curve. The maximum meaningful range is thus largest for points furthest from these boundaries and reduces to zero exactly at the turning points. Equation (12) approximates this phase dependence of the validity range with a $\cos^2$ function, and Equation (13) defines the maximum amplitude admissible to indeed never exceed turning points.

Combining the above results and reintroducing the detector pixel index j, the following expression for pixel (j) and phase step (i) wise phase stepping deviations $\Delta\phi_{ji}$ reducing (and, for sufficiently small $\Delta\phi_{ji}$, minimizing) the ℓ_2 error $\left(o_j + a_j \sin(\phi_i + \Delta\phi_{ji} - \phi_{0,j}) - y_{ji}\right)^2$ can be given, choosing $m_0 = \frac{1}{2}$:

$$\Delta\phi_{ji} \approx \begin{cases} 0 & a_j\cos(\phi_i - \phi_{0,j}) = 0 \\ \text{softlimit}\left(\dfrac{\left(\frac{y_{ji}-o_j}{a_j} - \sin(\phi_i - \phi_{0,j})\right)}{\cos(\phi_i - \phi_{0,j})}, \frac{1}{2}\cos^2(\phi_i - \phi_{0,j})\right) & \text{else} \end{cases}. \qquad (14)$$

Figure 3 shows an example of this approximate least squares solution to $\Delta\phi_{ji}$.

2.2.2. Noise Weighted Average of Phase Deviations

Now that an expression has been derived for the deviations $\Delta\phi_{ji}$ optimizing the abscissa values for individual PSCs indexed by j given previous sinusoid fits, the respective results for all PSCs sharing the same sampling locations may be averaged, such that the mean deviations $\Delta\phi_i$ essentially correspond to the actual systematic phase stepping errors (whereas statistical noise will mostly cancel out):

$$\Delta\phi_i = \frac{\sum_j w_{ji}\Delta\phi_{ji}}{\sum_j w_{ji}}, \qquad (15)$$

using weights w_{ji} factoring in the relative certainty and relevance of the individual $\Delta\phi_{ji}$. An appropriate choiceis

$$w_{ji} = a_j^2 \cos^2(\phi_i - \phi_{0,j}), \qquad (16)$$

where $\cos^2(\phi_i - \phi_{0,j})$ weights the slope dependent error propagation from noisy measurements y_{ji} to $\Delta\phi_{ji}$ based on the derivative of the sinusoid model at phase step i and a_j^2 weights the contribution of a particular PSC j to the accumulated ℓ_2 error. These considerations lead to

$$
\begin{aligned}
\Delta\phi_i &= \frac{\sum_j a_j^2 \cos^2(\phi_i - \phi_{0,j})\Delta\phi_{ji}}{\sum_j a_j^2 \cos^2(\phi_i - \phi_{0,j})} \\[2mm]
&= \frac{\sum_j a_j^2 \cos^2(\phi_i - \phi_{0,j}) \, \text{softlimit}\left(\frac{\left(\frac{y_{ji}-o_j}{a_j} - \sin(\phi_i-\phi_{0,j}) \right)}{\cos(\phi_i-\phi_{0,j})}, \frac{1}{2}\cos^2(\phi_i - \phi_{0,j}) \right)}{\sum_j a_j^2 \cos^2(\phi_i - \phi_{0,j})} \\[2mm]
&= \frac{\sum_j \text{softlimit}\left(\cos(\phi_i - \phi_{0,j}) \left(a_j \left(y_{ji} - o_j \right) - a_j^2 \sin(\phi_i - \phi_{0,j}) \right), \frac{1}{2}a_j^2 \cos^4(\phi_i - \phi_{0,j}) \right)}{\sum_j a_j^2 \cos^2(\phi_i - \phi_{0,j})} ,
\end{aligned}
\tag{17}
$$

where the last step uses the relation $\alpha \, \text{softlimit}(x, m) = \text{softlimit}(\alpha x, \alpha m)$ for arbitrary real-valued arguments x and m and factors $\alpha \geq 0$.

Finally, the above derivations can be combined to an iterative optimization algorithm reducing the accumulated least square error of multiple sinusoid fits (indexed by j) to data points y_{ji} over shared abscissa values ϕ_i as defined by Equation (6). A pseudo code representation is given in Algorithm 1, further introducing the relaxation parameter $\lambda_k \in (0; 1]$ that may be chosen <1 in order to damp the adaptions to $\phi_i^{(k)}$ if desired. The intermediate sinusoid fits may be accomplished using the iterative algorithm described in the previous section.

2.2.3. Inhomogeneous Sampling Phase Deviations

Up to now, it has been assumed that deviations from the intended phase stepping positions are due to purely translational uncertainties in the relative motion of the involved gratings, resulting in offsets $\Delta\phi_i$ of the actual from the intended sampling phases that are homogeneous throughout the whole detection area. When also considering relative grating period changes (e.g., due to either thermal expansion or motion induced changes in magnification) and rotary motions of the interferometer's gratings relative to each other (e.g., due to backlashes within the mechanical actuators), the effective sampling phases at each phase step may exhibit gradients over the detection area. Given the small grating periods (micrometer scale) compared to the total extents of the gratings (centimeter scale), both tilts in the sub-microrad range and relative period changes in the range of 10^{-7} will already manifest themselves in observable gradients.

The corresponding optimization problem regarding gradients is, analog to Equation (6), given by:

$$
o_j, a_j, \phi_{0,j}, \Delta\phi_i(j) = \underset{o_j, a_j, \phi_{0,j}, \Delta\phi_i(j)}{\text{argmin}} \sum_{i,j} \left(o_j + a_j \sin(\phi_i + \Delta\phi_i(j) - \phi_{0,j}) - y_{ji} \right)^2
\tag{18}
$$

where the spatial dependence of the phase deviations $\Delta\phi_i$ has been accounted for by a functional dependence on the detector pixel index j. Given the expected gradients (as illustrated in Figure 5), the spatially varying phase deviations $\Delta\phi_i(j)$ have the following form:

$$
\Delta\phi_i(j) = \Delta\phi_i + \nabla_h\phi_i\,(h - h_0) + \nabla_v\phi_i\,(v - v_0) + \nabla_{hv}\phi_i\,(h - h_0)(v - v_0) + \nabla_{h^2}\phi_i\,(h - h_0)^2,
\tag{19}
$$

with the coefficients $\nabla_h\phi_i$, $\nabla_v\phi_i$, $\nabla_{hv}\phi_i$ and $\nabla_{h^2}\phi_i$ quantifying the respective gradients in horizontal and vertical direction as well as the mixed term and the curvature in horizontal direction, and h and v

being spatial detector pixel indices related to the linear pixel index j through the amount N_h of pixels within one detector row:

$$j = vN_h + h$$
$$h = j \bmod N_h \tag{20}$$
$$v = (j - h)/N_h.$$

The constant offsets h_0 and v_0 characterize the detector center.

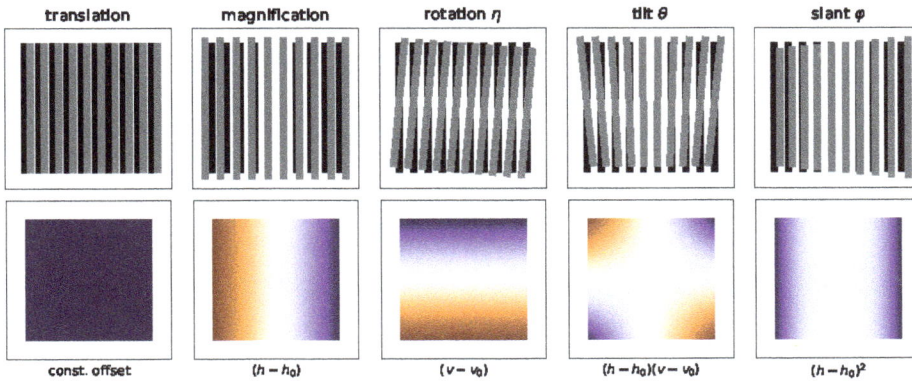

Figure 5. Grating misalignments (top row) and corresponding spatial phase variations (bottom row). From left to right: Translation, magnification, rotation, tilt, and slant. The latter two effects (as well as translation induced magnification) are only observable in cone beam setups. The employed color bar ranges from orange for negative values over white (zero) to blue for positive values.

The optimization of the extended objective function in Equation (18) can be performed analog to that of Equation (6) with the only difference lying in the evaluation of the spatial phase difference maps $\Delta\phi_{ji}$ defined by Equation (14) (an example of which is shown in Figure 6). The weighted average derived in the previous section to determine the homogeneous offset $\Delta\phi_i$ can be extended to a generalized linear least squares fit of the model $\Delta\phi_i(j) = \Delta\phi_i(h(j), v(j))$ defined by Equation (19) to the local estimates $\Delta\phi_{ji}$ (Equation (14), Figure 6), also taking the weights defined by Equation (16) into account. Said procedure is stated more formally in Algorithm 2.

Basic geometric considerations neglecting higher order interrelations of the considered effects (e.g., rotation and effective period change) result in the following relations between the observable parameters $\nabla_h\phi_i$, $\nabla_v\phi_i$, $\nabla_{hv}\phi_i$, $\nabla_{h^2}\phi_i$ and relative translatory and rotatory motions of the interferometer's gratings. To relate various magnification changes to spatial motions based on the intercept theorem, an assumption has to be made as to which of the gratings actually moved. Here, the grating that is mounted on the linear phase stepping actuator is assumed to be the cause of all relative motions of both gratings also including tilts and rotations. The "source–grating distance" in the following equations will thus refer to the stepped grating.

$\Delta\phi_i$ quantifies the translational error analog to Section 2.2.2. In contrast to the previous section, the present model distinguishes between homogeneous phase deviations induced by translation and the mean component induced by the $\nabla_{h^2}\phi_i(h - h_0)^2$ term in the case of non-vanishing curvature of the spatial phase deviation.

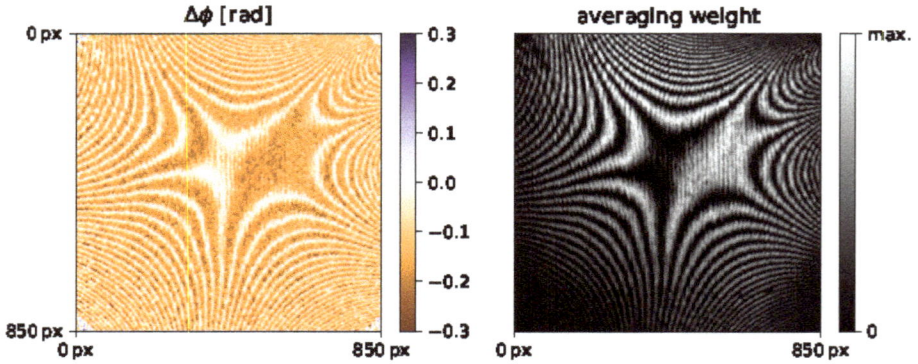

Figure 6. Example of sampling phase deviations (left) obtained from Equation 14 (cf. also Figure 3) for the first frame of the phase stepping series shown in Figure 1 with corresponding importance weights (**right**) as defined by Equation (16). White regions (0 on the colorbar) in the $\Delta\phi$ map (**left**) correspond to samples close to or exactly on turning points of the fitted sinusoids, where phase deviations cannot be effectively determined. These areas get little weighting in the determination of the average phase deviation as can be seen by the corresponding dark fringes in the weighting map on the right.

The vertical gradient parameter $\nabla_v\phi_i$ is related to a relative rotation η of both gratings about the optical axis:

$$\tan\eta = \frac{\nabla_v\phi_i}{2\pi}\frac{\text{effective grating period}}{\text{detector pixel pitch}}, \tag{21}$$

where the "effective grating period" refers to the projected period length at the location of the detector, which should be identical for both interferometer gratings (not considering the optional additional coherence grating close to the X-ray source).

The horizontal gradient parameter $\nabla_h\phi_i$ is related to a relative mismatch in effective grating periods of the gratings either due to relative distance changes along the optical axis or due to actual expansions (e.g., thermally induced):

$$\text{relative period mismatch} = \frac{\text{effective period difference}}{\text{effective grating period}} = \frac{\nabla_h\phi_i}{2\pi}\frac{\text{effective grating period}}{\text{detector pixel pitch}}. \tag{22}$$

When assuming relative grating period mismatches to be caused by changes in magnification due to translations of one of the gratings along the optical axis, the following relation applies to first order:

$$\begin{aligned}
\text{translation distance} &= (\text{relative period mismatch})\frac{(\text{source–grating distance})^2}{\text{source–detector distance}}\\
&= \frac{\nabla_h\phi_i}{2\pi}\frac{\text{effective grating period}}{\text{detector pixel pitch}}\frac{(\text{source–grating distance})^2}{\text{source–detector distance}}.
\end{aligned} \tag{23}$$

The change $\nabla_{hv}\phi_i$ of the horizontal gradient throughout the vertical direction corresponds to a relative change in magnification from top to bottom, e.g., due to a tilt θ of one of the gratings about the horizontal axis. Using the above relation between magnification changes and spatial displacements, the tilt θ about the horizontal axis is related to $\nabla_{hv}\phi_i$ approximately by

$$\tan\theta = \frac{\nabla_{hv}\phi_i}{2\pi} \frac{\text{effective grating period}}{\text{detector pixel pitch}} \left((\text{detector pixel pitch}) \frac{\text{source–grating distance}}{\text{source–detector distance}} \right)^{-1}$$

$$\times \frac{(\text{source–grating distance})^2}{\text{source–detector distance}} \quad (24)$$

$$= \frac{\nabla_{hv}\phi_i}{2\pi} \frac{(\text{effective grating period})(\text{source–grating distance})}{(\text{detector pixel pitch})^2}.$$

A non-vanishing curvature $\nabla_{h^2}\phi_i$ arises in the case of a rotary motion about the vertical axis (slant) and is analogously related to the slant angle φ to first order by

$$\tan\varphi = \frac{\nabla_{h^2}\phi_i}{2\pi} \frac{(\text{effective grating period})(\text{source–grating distance})}{(\text{detector pixel pitch})^2}. \quad (25)$$

2.3. Experimental Setup

The interferometer used to demonstrate the described analysis techniques consists of a set of three gratings (coherence grating G0, phase grating G1 and absorption grating G2) manufactured by microworks GmbH (Karlsruhe, Germany) for a design energy of 54 keV and mounted in an in house cone beam micro-CT setup comprising a commercial microfocus X-ray source operated at 80 kV acceleration voltage as well as a commercial flat panel detector with a pixel pitch of 74.8 µm. The grating periods are 4.8 µm (G0), 2.4 µm (G1) and 4.8 µm (G2), and the gratings are placed at 100 cm, 125 cm and 150 cm distance from the X-ray source, respectively. The field of view is limited by the diameter of G2 of approximately 10 cm. The G1 grating is mounted on a piezo driven linear actuator responsible for the phase stepping. The detector is placed right behind the absorption grating in about 155 cm distance from the source. Images are averaged over 10 exposures of 0.5 s each and are cropped to a region of $850 \times 850\,\text{px}^2$ corresponding to an area of $6.4 \times 6.4\,\text{cm}^2$ on the detector. A piece of plastic hose of 2 cm diameter placed about 143 cm distance from the source serves as the sample object.

3. Experiment and Results

Phase stepping series of 15 images sampled at varying relative grating shifts uniformly distributed over three grating periods have been acquired both with and without sample in the beam path. Figure 1 shows the first five frames of the empty beam series. The resulting phase stepping curves at each pixel (indexed by j) of the detector have been evaluated using a least squares fit to a sinusoid model parameterized by mean o_j, amplitude a_j and phase offset $\phi_{0,j}$ under the initial assumption of perfectly stepped gratings. These preliminary results are shown in Figure 2 and correspond to those obtained by classic Fourier analysis of the phase stepping curves. Deviations of the sampling positions from the intended ones are then determined based on systematic deviations of the sampled data from the fitted sinusoids by means of Equation (17) for all 15 frames of the phase stepping series. Figure 6 provides an example for the first frame of the series. By iterating the sinusoid fits and the corrections to the sampling positions to reduce the overall least squares error (cf. Equation (6)) by means of Algorithm 1, the sampling positions' deviations are found as shown in Figure 7. Figure 8 shows the reduction of Moiré modulated systematic errors in the final results, i.e., the transmission, visibility and differential phase images. The root mean square error is reduced by almost a factor of two in the present example and is already close to convergence after the first iteration as can be seen in Figure 9. Correspondingly, the deviations from the intended phase stepping positions are almost completely deduced within the first iteration, as shown in Figure 7 (bottom). Nevertheless, complete suppression of the Moiré artifacts within the final images requires some further iterations, as illustrated in Figure 10.

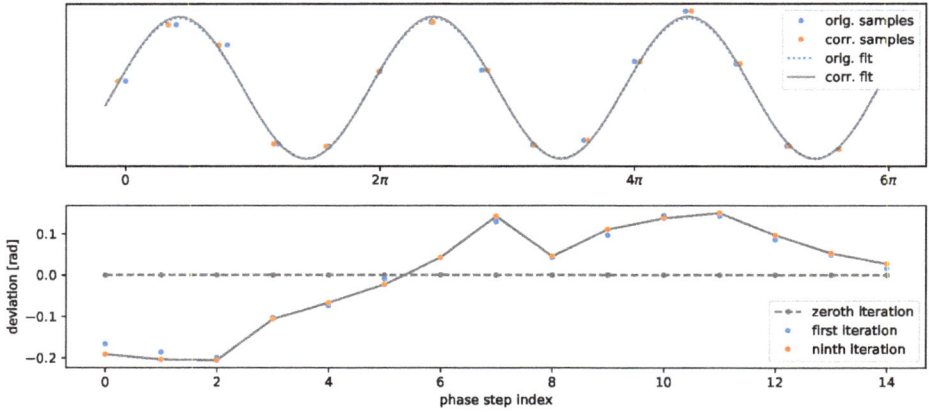

Figure 7. *Above:* An example phase stepping curve consisting of 15 steps over three grating periods. Sampled values are shown both at the originally intended as well as at the inferred sampling positions (blue and orange markers respectively) along with the corresponding initial and corrected sinusoid fits. Although the difference in the resulting fit appears small, it is clearly noticeable in the final images, as shown in Figure 8. *Below:* Deviations of the phase stepping series' sampling positions from the intended ones in units of radians as found at 0, 1 and 9 iterations of Algorithm 1. The range of deviations corresponds to roughly $\pm 10\%$ of the intended stepping increments of $\frac{2}{5}\pi$.

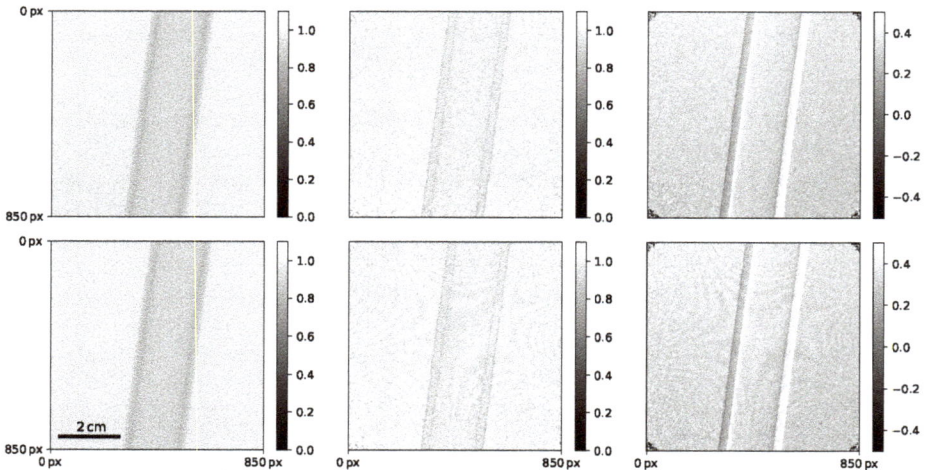

Figure 8. Transmission, visibility and phase images (from left to right) of the sample referenced to empty beam images. The top and bottom rows show results based on phase stepping curve evaluations with and without correction of the actual sampling positions, respectively. The evaluation based on the assumption of error free sampling positions (bottom row) exhibits distinctive systematic errors modulated by the Moiré structure of the reference phase image (cf. Figure 2).

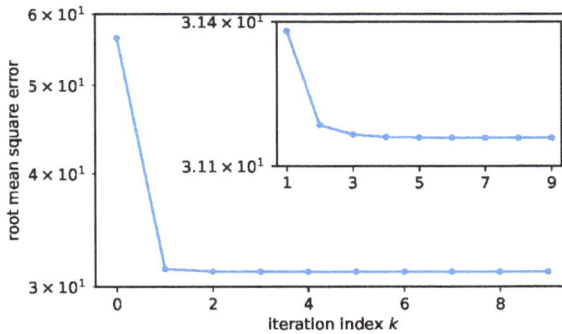

Figure 9. Root mean square error (RMSE) of the sinusoid fits to the empty beam phase stepping series throughout the iterations of Algorithm 1. After the first correction of the actual sampling locations by Equation (17), the error is reduced by almost a factor of two. The following iterations further reduce the error confirming the validity of Algorithm 1 for the solution of Equation (6). Additional consideration of spatially inhomogeneous stepping distances (Equation (2)) further reduces the RMSE by merely 0.1%.

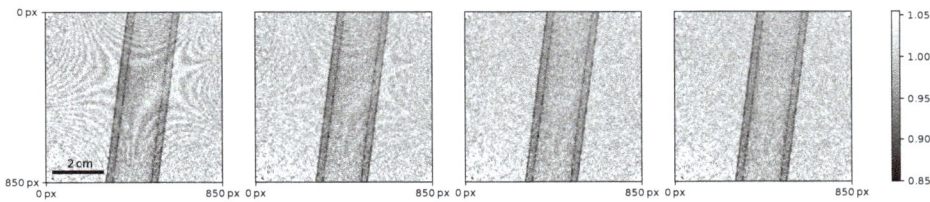

Figure 10. From left to right: Visibility images normalized to the respective empty beam images for 0 (regular analysis assuming perfect stepping), 1, 3 and 9 iterations of Algorithm 1 applied to the phase stepping series with and without sample, respectively. The grayscale window has been chosen to emphasize the Moiré artifacts.

In addition to the mean deviations of the phase steps from the intended positions, spatial gradients throughout the detection area have also been considered (cf. Algorithm 2). Figure 11 shows the respective differential deviations from the intended phase steps between the first nine frames of the phase stepping series, normalized to the nominal homogeneous phase stepping increment of $2\pi/5$. The mean contributions of each component are listed in Table 1. While the homogeneous error of 0.1 rad ranges within 10% of the nominal step size (or 2% of the grating period), the remaining effects are two to three orders of magnitude smaller. The root mean square error of the sinusoid fits for the whole phase stepping series is reduced by 0.1% relative to the optimization considering only homogeneous phase step deviations, as shown in Figure 9. Consequently, the derived images (not shown) are visually equivalent to those obtained previously (cf. Figure 8).

Figure 12 shows variations in the relative alignment of the gratings derived from the inhomogeneous phase stepping analysis by means of Equations (21)–(25). Besides deviations from the nominal linear motion of the gratings, minute rotations as well as subtle changes in relative magnification can also be detected. The correlation between grating rotations and translational errors visible in the left hand side graph in Figure 12 indicate rotations about an off-center pivot point about 10^{-1} m below the grating center, which is consistent with the actual placement of the phase stepping actuator in the experimental setup. Although the analog correlation between tilts about the horizontal axis and translation (along the optical axis) induced variations in magnification is much less pronounced (Figure 12, right hand side), the mean trend and magnitude are also consistent with the assumption of a pivot point below the field of view. However, the observed magnitude (10^{-7}) of the relative mismatch of the effective grating periods is well explicable by temperature variations in the order of magnitude of 10^{-1} K given a thermal expansion coefficient in the order of magnitude of 10^{-6} K^{-1} for the typical wafer materials silicon and graphite. Finally, the phase stepping inhomogeneities further suggest rotational motions about the vertical axis on the microrad scale (also shown in Figure 12).

As a crude error assessment, the standard error of the mean phase deviation can be estimated from the sinusoid fits' root mean square error:

$$\sigma_{\text{mean}} \approx \frac{1}{\sqrt{\text{contributing detector pixels}}} \frac{\text{sinusoid fit RMSE}}{\text{mean sinusoid amplitude}}$$

$$\approx \sqrt{\frac{2}{\text{detector pixels}}} \frac{\text{sinusoid fit RMSE}}{\text{mean sinusoid amplitude}} . \tag{26}$$

The latter corresponds for the present data set to about 6% of the mean observed sinusoid amplitude, which directly translates to 6×10^{-2} rad on the abscissa. Given the amount of detector pixels contributing to the least squares fits of $\Delta\phi_i(j)$ within each frame of the phase stepping series, a standard error in the order of magnitude of

$$\sigma_{\text{mean}} \approx 10^{-4} \, \text{rad} \tag{27}$$

results. This implies that, according to the results given in Table 1, the tilt and slant contributions $\nabla_{hv}\phi$ and $\nabla_{h2}\phi$ are close to the expected noise level for the present case.

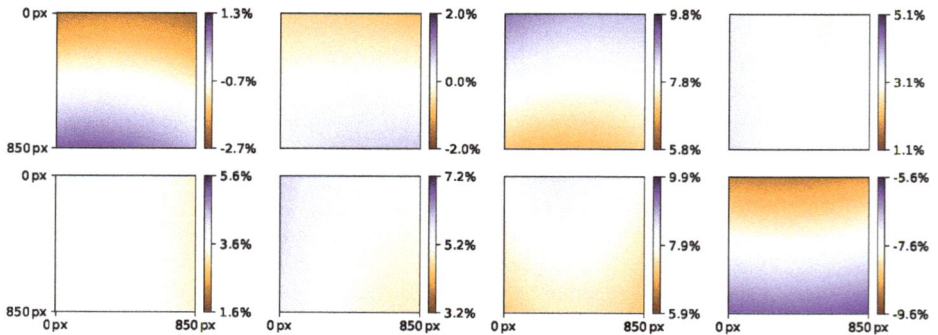

Figure 11. Relative deviations of the actual phase steps from the intended step width of $\frac{2\pi}{5}$ between the first nine frames of the phase stepping series ($\frac{5}{2\pi}(\Delta\phi_{i+1}(j) - \Delta\phi_i(j) - \frac{2\pi}{5})$). The variations $\Delta\phi_i(j)$ have been determined by optimization of Equation (18) assuming the spatial dependence defined by Equation (19) (see also Algorithm 2).

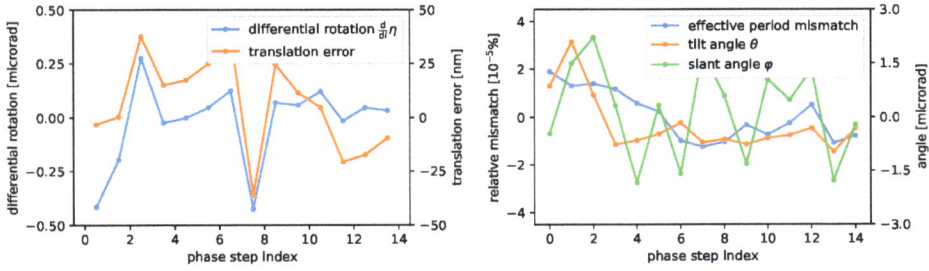

Figure 12. Quantitative results derivable from the inhomogeneities in the phase stepping deviations $\Delta\phi_i(j)$ (Equation (19)). On the left, the change in tilt angle about the optical axis from frame to frame within the phase stepping series is shown along with the accompanying linear motion error. Rotation correlated translations indicate an off-center pivot point. On the right hand side, the relative grating scaling error is shown along with the found tilt and slant about the horizontal and vertical axis. These quantities represent deviations from the mean grating alignment throughout the phase stepping series. The tilt and slant angles range close to the expected noise level (cf. Table 1 and Equation (27)).

Table 1. Root mean square contributions of the mean, gradient and curvature components of $\Delta\phi_i(j)$ to the sampling phase deviations found for the present phase stepping series in units of radians. The homogeneous error $\Delta\phi_i$ is by far the dominating effect. The contributions of $\nabla_{hv}\phi_i$ and $\nabla_{h^2}\phi_i$ range in the order of magnitude of the expected noise level of 10^{-4} rad (cf. Equations (26) and (27)).

$\sqrt{\overline{\Delta\phi_i^2}}$	$\sqrt{\overline{(\nabla_h\phi_i\,(h-h_0))^2}}$	$\sqrt{\overline{(\nabla_v\phi_i\,(v-v_0))^2}}$	$\sqrt{\overline{(\nabla_{hv}\phi_i\,(h-h_0)(v-v_0))^2}}$	$\sqrt{\overline{(\nabla_{h^2}\phi_i\,(h-h_0)^2)^2}}$
1.2×10^{-1}	2.4×10^{-3}	3.6×10^{-3}	3.6×10^{-4}	7.5×10^{-4}

4. Discussion and Conclusions

A fast converging iterative algorithm for the joint optimization of both the sinusoid model parameters and the actual sampling locations for the evaluation of grating interferometric phase stepping series has been proposed. The additional effort (with respect to [5,6]) of explicitly optimizing phases rather than Fourier coefficients allows for a straight forward extension of the optimization algorithm also in the case of spatially varying phase stepping increments due to further mechanical degrees of freedom besides the intended translatory stepping motion. By division of the full optimization task into three easily tractable subproblems (phase stepping curve fitting, identification of sampling position deviations and fitting of the latter to an expected spatial model), the use of generic nonlinear optimization algorithms as, e.g., used in [7] is avoided. Of these subtasks, only the second one is actually nonlinear, and a significant portion of the present article has been devoted to its approximate linearization (also taking noise propagation into consideration). The problem of simultaneous phase stepping curve evaluation and (spatially varying) sampling position determination is thus reduced to the iterative alternation of two generalized linear least squares problems and one linear approximation to a nonlinear problem. Due to this almost linear nature, sufficient convergence is achieved within less than five iterations for the presented example. Although no rigorous convergence analysis has been performed, the outlined structure of the optimization problem does not raise severe concerns regarding its stability. In cases of doubt, convergence can simply be slowed down by means of the relaxation parameter λ_k.

For the presented example data set, mean phase stepping errors of up to 10% of the nominal step width have been found, and their correction results in both a considerable reduction of the overall root mean square error by almost a factor of two and a significant visual improvement of the final images. Although higher order effects are observable, their contribution was found to be two to three orders of magnitude smaller than that of the mean stepping error, and their correction thus did not contribute to further improvements in visual image quality in the present case. However, the higher order deviations allow the detection of minute motions of the gratings and thus provide a valuable tool for the monitoring and debugging of experimental setups. First order approximations for the relations between spatial phase variations and mechanical degrees of freedom of the moved grating have been given. For the present data set, linear motion errors up to 25 nm as well as rotational motions on the microrad scale have been inferred from the phase stepping series. While the tilt and slant angles about the horizontal and vertical axes, respectively, have been found to range close to the expected noise level and should rather be interpreted as upper limits to actual motions, magnification changes in the range of 10^{-7} and sub-microrad rotations about the optical axis were well detectable. The expected correlations between rotation and translation due to an off-center pivot point further support the plausibility of the results. This crosstalk between sub-microrad rotations and effective translations further indicates that noticeable phase stepping errors will be almost inevitable even for very carefully designed experiments, wherefore an optimization based evaluation of the phase stepping series as proposed in Algorithm 1 is generally advisable. With processing speeds in the range of 0.1 s per phase stepping series (using graphics processors), it is well suitable as a standard processing method also for large image series.

Author Contributions: Conceptualization, J.D., A.B. and S.Z.; Methodology, J.D.; Software, J.D.; Validation, J.D.; Formal Analysis, J.D.; Investigation, J.D. and A.B.; Resources, S.Z. and A.B.; Data Curation, J.D.; Writing—Original Draft Preparation, J.D.; Writing—Review and Editing, J.D., A.B. and S.Z.; Visualization, J.D.; Supervision, S.Z.; and Project Administration, S.Z.

Funding: The research was funded by the project group "Nano-CT Systems" (NCTS) of the Fraunhofer IIS/EZRT, which was supported by the Bavarian State ministry of Economic Affairs, Infrastructure, Transport and Technology.

Acknowledgments: The authors acknowledge R. Hanke for facilitating the present work, K. Dremel and D. Althoff for operation of the utilized in house X-ray setup and A. Haydock for proof-reading the manuscript with respect to English language. Further, the scipy libraries (http://www.scipy.org) are acknowledged, which were extensively used to perform the presented analyses.

Conflicts of Interest: The authors declare no conflict of interest.

Abbreviations

The following abbreviations are used in this manuscript:

PSC   Phase stepping curve
RMSE   Root mean square error

References

1. David, C.; Nöhammer, B.; Solak, H.H.; Ziegler, E. Differential x-ray phase contrast imaging using a shearing interferometer. *Appl. Phys. Lett.* **2002**, *81*, 3287–3289. [CrossRef]
2. Momose, A.; Kawamoto, S.; Koyama, I.; Hamaishi, Y.; Takai, K.; Suzuki, Y. Demonstration of X-Ray Talbot Interferometry. *Jpn. J. Appl. Phys.* **2003**, *42*, L866–L868. [CrossRef]
3. Pfeiffer, F.; Weitkamp, T.; Bunk, O.; David, C. Phase retrieval and differential phase-contrast imaging with low-brilliance X-ray sources. *Nat. Phys.* **2006**, *2*, 258–261. [CrossRef]
4. Revol, V.; Kottler, C.; Kaufmann, R.; Straumann, U.; Urban, C. Noise analysis of grating-based x-ray differential phase contrast imaging. *Rev. Sci. Instrum.* **2010**, *81*, 073709. [CrossRef] [PubMed]
5. Seifert, M.; Kaeppler, S.; Hauke, C.; Horn, F.; Pelzer, G.; Rieger, J.; Michel, T.; Riess, C.; Anton, G. Optimisation of image reconstruction for phase-contrast x-ray Talbot–Lau imaging with regard to mechanical robustness. *Phys. Med. Biol.* **2016**, *61*, 6441–6464. [CrossRef] [PubMed]
6. Vargas, J.; Sorzano, C.O.S.; Estrada, J.C.; Carazo, J.M. Generalization of the Principal Component Analysis algorithm for interferometry. *Opt. Commun.* **2013**, *286*, 130–134. [CrossRef]
7. Kaeppler, S.; Rieger, J.; Pelzer, G.; Horn, F.; Michel, T.; Maier, A.; Anton, G.; Riess, C. Improved reconstruction of phase-stepping data for Talbot-Lau x-ray imaging. *J. Med. Imaging* **2017**, *4*, 034005. [CrossRef] [PubMed]

8. Von Teuffenbach, M.; Noichl, W.; Brendel, B.; Herzen, J.; Pfeiffer, F.; Koehler, T.; Noël, P.B. Grating Position Estimation for Grating-based Computed Tomography. In Proceedings of the Conference on X-ray and Neutron Phase Imaging with Gratings, XNPIG 2017, Zurich, Switzerland, 12–15 September 2017; pp. 169–170.
9. Ritter, A.; Anton, G.; Weber, T. Simultaneous Maximum-Likelihood Reconstruction of Absorption Coefficient, Reconstruction of Absorption Coefficient, Refractive Index and Dark-Field Scattering Coefficient in X-Ray Talbot-Lau Tomography. *PLoS ONE* **2016**, *11*, e0163016. [CrossRef] [PubMed]
10. Marco, F.D.; Marschner, M.; Birnbacher, L.; Noël, P.; Herzen, J.; Pfeiffer, F. Analysis and correction of bias induced by phase stepping jitter in grating-based X-ray phase-contrast imaging. *Opt. Express* **2018**, *26*, 12707–12722. [CrossRef] [PubMed]
11. Yashiro, W.; Terui, Y.; Kawabata, K.; Momose, A. On the origin of visibility contrast in x-ray Talbot interferometry. *Opt. Express* **2010**, *18*, 16890–16901. [CrossRef] [PubMed]
12. Bech, M.; Bunk, O.; Donath, T.; Feidenhans'l, R.; David, C.; Pfeiffer, F. Quantitative x-ray dark-field computed tomography. *Phys. Med. Biol.* **2010**, *55*, 5529–5539. [CrossRef] [PubMed]
13. Strobl, M. General solution for quantitative dark-field contrast imaging with grating interferometers. *Sci. Rep.* **2014**, *4*, 7243. [CrossRef] [PubMed]
14. Jensen, T.H.; Bech, M.; Bunk, O.; Donath, T.; David, C.; Feidenhans'l, R.; Pfeiffer, F. Directional x-ray dark-field imaging. *Phys. Med. Biol.* **2010**, *55*, 3317–323. [CrossRef] [PubMed]
15. Revol, V.; Kottler, C.; Kaufmann, R.; Neels, A.; Dommann, A. Orientation-selective X-ray dark field imaging of ordered systems. *J. Appl. Phys.* **2012**, *112*, 114903. [CrossRef]
16. Malecki, A.; Potdevin, G.; Biernath, T.; Eggl, E.; Willer, K.; Lasser, T.; Maisenbacher, J.; Gibmeier, J.; Wanner, A.; Pfeiffer, F. X-ray tensor tomography. *Europhys. Lett.* **2014**, *105*, 38002. [CrossRef]
17. Bayer, F.; Hu, S.; Maier, A.; Weber, T.; Anton, G.; Michel, T.; Riess, C. Reconstruction of scalar and vectorial components in X-ray dark-field tomography. *Proc. Natl. Acad. Sci. USA* **2014**, *111*, 12699–12704. [CrossRef] [PubMed]
18. Vogel, J.; Schaff, F.; Fehringer, A.; Jud, C.; Wieczorek, M.; Pfeiffer, F.; Lasser, T. Constrained X-ray tensor tomography reconstruction. *Opt. Express* **2015**, *23*, 15134–15151. [CrossRef] [PubMed]
19. Wieczorek, M.; Schaff, F.; Pfeiffer, F.; Lasser, T. Anisotropic X-ray Dark-Field Tomography: A Continuous Model and its Discretization. *Phys. Rev. Lett.* **2016**, *117*, 158101. [CrossRef] [PubMed]
20. Sharma, Y.; Wieczorek, M.; Schaff, F.; Seyyedi, S.; Prade, F.; Pfeiffer, F. Six dimensional X-ray Tensor Tomography with a compact laboratory setup. *Appl. Phys. Lett.* **2016**, *109*, 134102. [CrossRef]
21. Dittmann, J.; Zabler, S.; Hanke, R. Nested Tomography: Application to Direct Ellipsoid Reconstruction in Anisotropic Darkfield Tomography. In Proceedings of the Conference on X-ray and Neutron Phase Imaging with Gratings, XNPIG 2017, Zurich, Switzerland, 12–15 September 2017; pp. 49–50.

Journal of
Imaging

MDPI

Article

Imaging with a Commercial Electron Backscatter Diffraction (EBSD) Camera in a Scanning Electron Microscope: A Review

Nicolas Brodusch [1,*], Hendrix Demers [2] and Raynald Gauvin [1]

[1]   Department of Mining and Materials Engineering, McGill University, Montréal, QC H3A 0C5, Canada; raynald.gauvin@mcgill.ca

[2]   Institut de Recherche d'Hydro-Québec, 1806 Boulevard Lionel-Boulet, Varennes, QC J3X 1S1, Canada; demers.hendrix@ireq.ca

*   Correspondence: nicolas.brodusch@mcgill.ca; Tel.: +1-51-4398-7182; Fax: +1-51-4398-4492

Received: 21 May 2018; Accepted: 20 June 2018; Published: 1 July 2018

Abstract: Scanning electron microscopy is widespread in field of material science and research, especially because of its high surface sensitivity due to the increased interactions of electrons with the target material's atoms compared to X-ray-oriented methods. Among the available techniques in scanning electron microscopy (SEM), electron backscatter diffraction (EBSD) is used to gather information regarding the crystallinity and the chemistry of crystalline and amorphous regions of a specimen. When post-processing the diffraction patterns or the image captured by the EBSD detector screen which was obtained in this manner, specific imaging contrasts are generated and can be used to understand some of the mechanisms involved in several imaging modes. In this manuscript, we reviewed the benefits of this procedure regarding topographic, compositional, diffraction, and magnetic domain contrasts. This work shows preliminary and encouraging results regarding the non-conventional use of the EBSD detector. The method is becoming viable with the advent of new EBSD camera technologies, allowing acquisition speed close to imaging rates. This method, named dark-field electron backscatter diffraction imaging, is described in detail, and several application examples are given in reflection as well as in transmission modes.

Keywords: dark-field (DF); electron channeling contrast imaging (ECCI); electron channeling pattern (ECP); electron backscatter diffraction (EBSD); deformation; scanning electron microscope (SEM); transmission kikuchi diffraction (TKD; t-EFSD; t-EBSD)

1. Introduction

Among the characterization tools available to the materials scientist, the scanning electron microscope (SEM) is probably the most used and versatile. It offers a spatial resolution in the sub-nanometer level when equipped with a cold-field emitter or a beam monochromator [1,2], and its design allows the analyst to observe the surface of a specimen from the nanometer to the centimeter range. Because the mean free path of low energy electrons is dramatically smaller than that of X-rays, the technique is the missing link between high penetration depth techniques like X-ray-based systems and atomic level surface techniques like atomic force microscopy or scanning tunneling microscopy. Thus, by varying the accelerating voltage (E_0) applied to the electron beam and the type of signal collected, one can obtain "bulk" information; however, if lower voltages are used, the collected signals originate from the shallow surface layers [2]. For example, the range of electrons/material interactions at E_0 = 20 kV in iron is roughly 1 μm, while it falls to 15 nm at E_0 = 1 kV. Note that these numbers are maximum values and that the emission depth and lateral distribution may be further reduced depending on the type of signal that is collected to generate the image.

Technically speaking, in a SEM, a beam of focused electrons is scanned over a specimen surface in a raster fashion, and many different signals are collected to generate different images with various contrasts. These contrasts depend on the nature of the particle that is collected, as well as on the nature of the interaction that the particle has undergone during its path through the material. Secondary electrons are produced by the ejection of mostly valence electrons due to atomic ionization, and are characterized by their low energy and small inelastic mean free path, in addition to being highly absorbed before reaching the exit surface. Their emission depth is thus confined to surface layers—generally in the range of a few nanometers—depending on the atoms' band structures and absorption, and they provide topographic contrast that allows us to observe the relief of the surface in a 3D-like fashion.

In contrast, the primary electrons that are backscattered towards the surface due to the atoms' Coulomb attraction forces retain sufficient energy to reach the exit surface with limited absorption, and carry information about the composition of the volume of material "seen" by these electrons. These are called Backscattered Electrons (BSE), and are responsible for compositional contrast (also known as material or Z contrast), which is in fact related to the mean atomic number of the material interacting with them. Their emission depth and lateral distribution depend on the material characteristics and the primary beam accelerating voltage. The energy distribution of these BSEs being material and SEM parameters dependent, energy filtration allows us to collect only high energy BSEs, i.e., those with low-loss of energy, which are associated with high spatial resolution and reduced interaction volume. These low-loss electrons suffer a small number of interactions, and originate from the close surrounding of the beam impact point on the surface. The depth resolution is of the order of the mean free path, but depends mostly on the energy-loss considered.

Because they interact with the crystal lattice of the specimen through diffraction processes, the BSEs carry information about the crystallinity in their emission volume, but mostly from the exit surface. In the SEM, there are two ways of gathering the diffraction information carried by these electrons. In 1967, Coates evidenced the channeling of BSEs by imaging a Kikuchi-like patterns from Ge and GaAs crystals when the electron beam was scanned over the surface of the specimens at low magnification [3]. This type of pattern was later termed as an Electron Channeling Pattern (ECP); this led to the well-known electron channeling contrast imaging (ECCI). Later, in 1973, Venables and Hartland obtained similar patterns in spot mode when the specimen was highly tilted towards a phosphorescent screen, typically 60–70° [4]. At this time, the image was captured from a phosphorescent screen by means of an external camera. Currently, charge-coupled device (CCD) cameras are commonly used to capture these patterns with high speed rates [5]—up to 3000 patterns per second—to produce phase and orientation maps based on the crystallography and crystallographic orientation of the different phases present in the analyzed material.

This discovery resulted in one of the most important techniques in materials science, namely Electron Backscattered Diffraction (EBSD) [6], that is now widely used in the materials community to characterize microstructures at the sub-micron scale with spatial resolution roughly ranging from 20 to 150 nm, depending upon the material's atomic number and density [7,8]. In this technique, the bands detected on the EBSD patterns (EBSP) are a projection of the crystal planes on the EBSD camera screen. They are processed and compared to a look-up table to match the most probable phase and orientation at each pixel. However, it took many years before the raw signal collected by the CCD cameras was used to generate images, the information gathered so far being mainly related to the bands detected on the EBSD pattern. In parallel, Prior and co-workers reported compositional and crystallographic orientation imaging when attaching solid state diodes just below and above the CCD screen [9]. They described a dramatic change in contrast when switching from the bottom to the top diodes to record the image. Top diodes provided compositional contrast, while those from the bottom resulted in orientation contrast. These findings were later confirmed and used by Payton et al. to help with phase identification when combined with EBSD indexing [10].

Recently, the use of the raw signal collected by the CCD camera was reported by several authors. In 2006, Wells and co-workers reported the first use of the EBSD camera as an imaging device, and more importantly, as a BSE angular distribution collector [11]. Following this route, Schwarzer and Sukkau demonstrated that imaging was possible by summing pixels from a specified region of the CCD camera and reconstructing the image based on the summed intensities [12,13], thus resulting in different contrasts depending on the pixels' location, as later confirmed by Nowell and co-workers [14]. In 2015, Wright and co-workers reported the first commercial software proposing this facility in a live scanning mode or in post-processed mode [15]. However, the images thus obtained did not receive the attention required to properly understand the contrasts that were observed at that time. At the same time, Brodusch and co-workers reported a post-processed imaging technique based on EBSPs used to generate dark-field type images related to the specific diffraction conditions selected on the diffraction pattern [16,17]. They named this technique EBSD dark-field imaging mode (EBSD-DF). More recently, and following the work of Wells, De Graef and co-workers reported a method based on EBSPs to measure surface topography via the determination of surface normals [18], and reported nanometer scale spatial resolution comparable with atomic force microscopy on a Ni surface machined with a femtosecond laser [19].

In this contribution, we report on how to use the CCD camera of a commercial EBSD system as an imaging detector in various situations, and how relevant it is to understand and optimize the image contrasts. First, the method will be described in details and a program developed in Python language will be provided for the reprocessing of post-acquisition EBSD data. The usefulness of our approach will be demonstrated by applying this method to materials providing compositional, magnetic domain and diffraction/Z-contrasts.

2. Materials and Methods

2.1. Scanning Electron Microscopy

In this work, imaging and EBSD works were accomplished using a Hitachi SU8000 cold-field emission scanning electron microscope (Hitachi High-Tech, Tokyo, Japan) equipped with an Oxford Instruments EBSD system. The EBSD camera was a Nordlys II with dimensions of 28×38 mm^2 and was controlled by the Flamenco software, part of the HKL Channel 5 suite. The energy cut-off of the camera was estimated to be not higher than 1 keV from previous measurements [20]. The accelerating voltage was 30 kV, except when otherwise specified, and the specimens were tilted to 70° toward the EBSD camera, except for the transmission EBSD set-up where the tilt was −20°. Parameters for each EBSD scan are given in the result section when necessary. The acquisition of the ECP in was performed with a Hitachi SU3500 variable-pressure SEM using a solid-state BSE detector at $E_0 = 30$ kV.

The X-ray images were obtained using a Bruker FlatQuad annular silicon drift detector (Bruker, Billerica, MA, USA) for energy dispersive spectrometer (EDS) attached to a Hitachi SU8230 cold-field emission scanning electron microscope. The accelerating voltage was 5 kV. Secondary electron (SE) images were acquired with an in-chamber Everhart-Thornley type detector providing BSE contrast due to the high number of SEs of type II and III (SEs generated by the BSEs inside and outside the specimen, respectively) collected by this detector.

For convention, the angle between the electron beam and the specimen surface normal directions is referred to θ_{in} while the angle made by the emitted BSEs with the specimen surface is referred to θ_{out}, as indicated in Figure 1a. In transmission mode, θ_{out} is referred to the angle between the transmitted electrons emission angle and the beam direction (Figure 1b). Hence, in this work, θ_{in} was 70° and −20° for conventional and transmission EBSD set-ups, respectively. θ_{out} may be calculated by simple geometric manipulations, and an example of a calibration curve relating the distance from the bottom of the EBSD camera screen with the corresponding θ_{out} is given in Figure 1c.

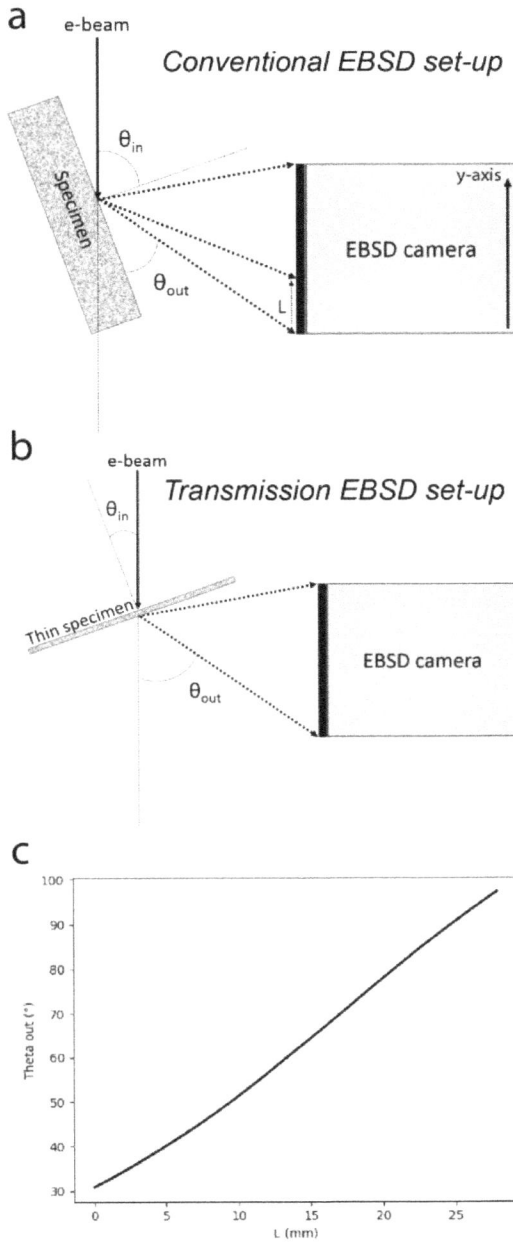

Figure 1. Schematics describing (**a**) the conventional electron backscattered diffraction (EBSD) set-up and (**b**) the transmission EBSD (t-EBSD) set-up; (**c**) Calibration relating the BSE emission angle θ_{out} and the distance L from the bottom of the EBSD camera screen. SINGLE COLUMN.

Bright-field (BF) images were recorded using the transmission mode of the SEM (STEM) with an yttrium aluminum garnet (YAG) scintillator detector located below the specimen around the optic axis of the microscope. An aperture was placed on top of the detector to select a BF collection angle of 10–20 mrad. The accelerating voltage was 30 kV. Note that the homemade sample holder used for this work allowed us to collect BF-STEM images and transmission EBSD (t-EBSD) data without any change of geometry or detector position, as described in a previous paper [21,22].

Monte Carlo simulations were conducted following the single scattering model as described by Gauvin and co-workers [23] using the screened Rutherford elastic cross-section expression and the Bethe continuous energy-loss equation, as modified by Joy and Luo [24]. The magnetic deflection was implemented following the procedure described by Newbury [25,26].

The image contrast C between regions A and B of an image was calculated using the following equation:

$$C_{A-B} = \frac{I_A - I_B}{I_A + I_B}$$ (1)

where I_A and I_B were the integrated intensities in area A and B, respectively.

In the case of magnetic domain contrast imaging (Section 3.3), a correction had to be applied to reduce the image tip noise generated by the cold-field emitter of the microscope [2]. This noise was observable because high gain needed to be applied to make the domain structure visible, as magnetic domain contrast with E_0 = 30 kV and a tilt angle of 60° lies between 0.15% and 0.20%, as reported by Newbury [27]. This noise appears as intensity changes from line to line all along the captured image. The post-processing correction was accomplished with a simple procedure using a code written in Python language (www.python.org). For each line of the image, the mean intensity was calculated and compared to the previous line, the first line being the reference mean intensity. Note that any line could be selected as the reference if desired. Then, the brightness of the line was adjusted to fit its mean intensity to that of the previous line and the adjusted brightness factor was finally applied to each pixel of the line. This procedure allowed us to reduce greatly the tip noise observed although not completely.

2.2. Materials

The material used for the acquisition of the ECP and EBSP displayed in Figure 2 was a 1×1 cm^2 (001) oriented LaAlO$_3$ single crystal provided by Alfa Aesar (Haverhill, MA, USA) (www.alfa.com). The application examples presented here were obtained on three different samples. The first was a piece of slag left over after melting lead-rich ores by early civilizations from Mexico. It was composed of mixed oxide and sulfide zinc/iron/lead-rich phases. The second sample was a non-oriented Fe-2.78%-Si electrical steel (NOES) on which magnetic domains were already observed and studied in a previous work [28]. These two samples were prepared by mechanical grinding and polishing down to 50 nm colloidal suspension, followed by ion milling with an accelerating voltage of 5 kV and an incident angle of 7° relative to the specimen surface. For this purpose, a Hitachi IM3000 (Hitachi High-Tech, Tokyo, Japan) flat milling system was used. The last sample was an AA2099 Al-Li-Cu alloy thinned to electron transparency by means of twin-jet electropolishing followed by ion milling at 2 kV accelerating voltage and 7° incidence angle on both sides of the foil.

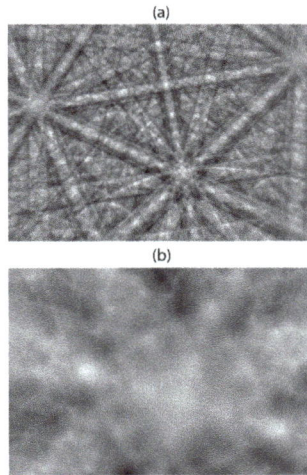

Figure 2. Typical pseudo-Kikuchi pattern obtained in a scanning electron microscope with (**a**) the electron backscattered diffraction (EBSP) technique and (**b**) the electron channeling technique (ECP) on a (001) oriented LaAlO$_3$ single crystal. Accelerating voltage was 20 kV in (**a**) and 30 kV in (**b**). Note the level of detail available in the (001) zone axis displayed in (**b**). SINGLE COLUMN.

2.3. Description of the Method

A full description of the method is available elsewhere [17], and only a short description is given here. The original method was based on the relation between the ECP and the EBSP, which are related by the accepted reciprocity theorem [29,30]. This relation explains that the two techniques should give identical patterns if the sources and detectors are switched when passing from one technique to the other. For sake of comparison, an EBSP and an ECP obtained from a (001) oriented LaAlO$_3$ single crystal are shown in Figure 2a,b, respectively. Note the higher magnification of the ECP (Figure 2b), which is due to the limited SEM scanning angle available to generate the pattern. The angular resolution of the two recorded patterns, defined as the acquisition angular step, can be estimated by dividing the total collection angle by the number of pixels of the image x-axis. For this reason, the ECP exhibits an improved angular resolution compared to the EBSP, as recently demonstrated by Brodusch and co-workers in their Figure 2 [17]. However, if the EBSD detector distance is increased, the angular resolution becomes comparable, and the main features of the ECP are well reproduced in the EBSP. A comparison of orientation images obtained from ECCI and the corresponding image using the reconstruction method based on EBSP was reported by Kaboli and co-workers [31]. They demonstrated matching contrasts between the two images, confirming the equivalent information contained in the two types of patterns.

The method originally described by Brodusch et al. was inspired by the contrast mechanism of ECCI. In ECCI, the intensity at each pixel of the image is determined by the intensity at the center of the ECP, which corresponds to the optic axis of the SEM [32]. When the local orientation changes, a shift of the ECP is observed, and the intensity at the optic axis point of the ECP, modulated by its angular resolution, changes. In our technique, a dataset of EBSPs is first acquired and stored. Then, a reference EBSP is chosen from a specific location in the original EBSD map and an array of pixels or a single pixel is selected on that reference pattern. The intensity is summed over the reference EBSP's pixel array and assigned to the pixel in the reconstructed map (EBSD map). This procedure is repeated for each pixel of the original map, and the resulting reconstructed map is named EBSD-DF image. Thus, the pixel array in the reference image acts as a virtual beam or a virtual aperture similar to that used for dark-field imaging in the TEM. The procedure described above is illustrated in Figure 3. Note that for convention, the term EBSP will be used throughout the text as a reference to the image collected by the EBSD camera, even if it does not contain any Kikuchi lines. It is important to note that our method gathers a lot of information on a specific sample area with only one scan, but is mostly intended to help the SEM user to optimize the collection angles of the BSE detector (the forward scatter detector, FSD, in highly tilted condition) and to position the EBSD camera accurately to maximize the image contrasts obtained with the FSD detector.

A program capable of reprocessing the EBSD dataset was developed, based on the Python programming language (https://python.org), and is available via the repository hosting service platform Bitbucket (https://bitbucket.org) at the address https://bitbucket.org/brodusch/py_ebsd_df. To accelerate the access and processing of the dataset, the stored EBSPs are first transferred into an hdf5 file (https://support.hdfgroup.org) in the form of a datacube. To generate the EBSD-DF images, the array of pixels is selected by the user, and EBSD-DF images can be saved if necessary.

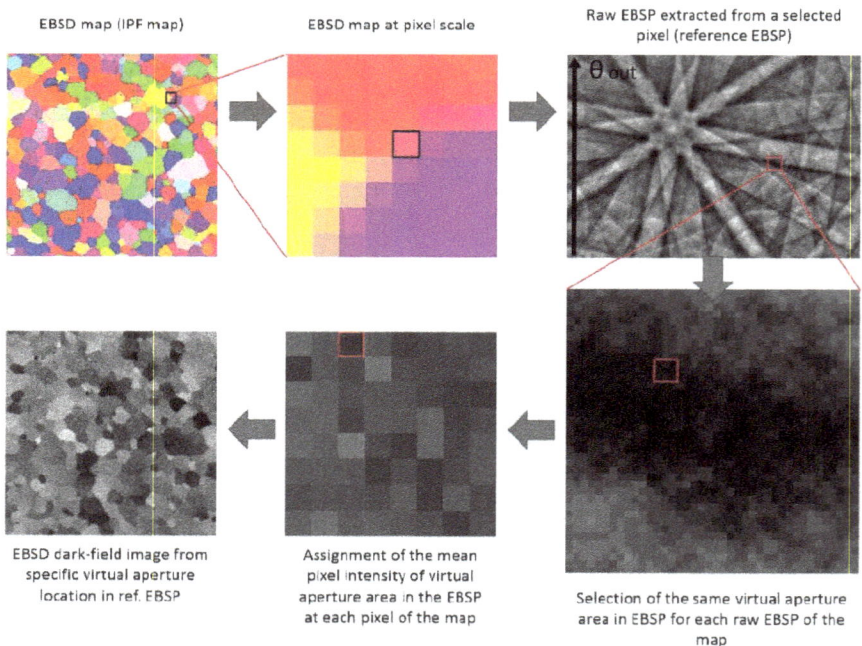

Figure 3. Description of the procedure to obtain post-processed EBSD-DF images from a stored EBSD dataset. DOUBLE COLUMN.

3. Results

3.1. EBSD-Dark-Field Imaging

Originally, the EBSD-DF method was developed to generate images for which the contrast was clearly related to the diffraction reflection selected [16,17], thus generating a large number of ECCI-like images from a single scan. This application essentially holds for single orientation areas where only slight deviations from the selected diffraction condition will produce interpretable contrast, similarly to ECCI. This resulted in a better description of the microstructure of the specimen, especially where complex deformation structures were expected, like those surrounding micro- and nano-indents. It has to be noted that in order to obtain a similar set of images from conventional ECCI analyses, one would need to rotate and tilt the specimen (or the beam) to generate a new image for each specific diffraction condition, and this, hundreds of times. This would be time consuming and is not realistic.

In addition to this, we demonstrated the capability of this new technique to simulate ECCI contrast with known conditions to improve the understanding of the contrasts observed when performing an ECCI analysis. The similarity between the ECCI and EBSD-DF images as shown in [31] confirmed that the primary assumption at the origin of the method was appropriate, and that the results obtained by applying the EBSD-DF technique could be transferred to ECCI. Work is in progress on this subject, and will be reported in a separate publication.

3.2. Compositional Imaging

It is known that the collection angle sustained by a BSE detector influences the contrast of the resulting BSE image, as recently reported by Aoyama and co-workers [33,34]. The raw signal collected by an EBSD detector reflects the angular distribution of the backscattered and forwarded electrons from a highly tilted specimen under electron bombardment. Wells et al. [11] first reported the impact of the collection area on the image obtained when the EBSD detector was used to generate an image. Here we used a specimen containing several distinct phases to evidence the main differences in terms of contrast regarding the position of the pixel array on the EBSD screen plane.

In order to study the impact of the angular distribution of BSEs on compositional contrast, a sample consisting of several mineral phases was investigated. To describe the chemistry of the sample, qualitative EDS background subtracted (net intensity) X-ray images were recorded, and are displayed in Figure 4 for elements S, Fe, Zn, and Pb. As shown, this specimen was mainly composed of sulfide phases. The dark areas in all maps at the same location were considered to be porosity.

Figure 5 presents reconstructed images of the same sample but from a different area, with specific collection areas in the reference image using the procedure described in Section 2.3. For each image, the inset shows the reference image captured via the EBSD detector with the collection area defined by the white square. Due to the limited detector distance achievable with our system, the BSE angular distribution was not totally captured by the EBSD camera. However, the extent of angular distribution collected by this system still allows one to draw useful conclusions. The image displayed in Figure 5a was obtained when the intensity from the full area of the camera screen was collected, i.e., without angular filtration. The contrast observed is typical of compositional contrast, and the major phases identified in Figure 4 are easily recognized. When only a small pixel array around the maximum of intensity of the reference image was selected (Figure 5b), a loss of contrast was noted, although it did not prevent the different phases from being properly identified. However, when comparing images obtained by collecting pixel arrays from the top (Figure 5c) and bottom (Figure 5d) center areas of the reference image, the change in contrast was more striking. Indeed, the contrast obtained with BSEs emitted towards the top of the EBSD camera screen, i.e., with high θ_{out} angles, produced a dramatic increase of compositional contrast, while the image obtained by collecting low θ_{out} exit angles exhibited a very weak compositional contrast. However, when low emission angles (θ_{out}) were collected (Figure 5d), the surface topography became more pronounced, eventually hiding compositional details such as those inside the large pore observed in the center bottom left of the image. The topography observed at the surface of this specimen was expected, and is an artefact of the flat ion milling technique that was applied to the specimen prior to the EBSD acquisition. In addition, a contrast inversion can be noticed between the dark (pyrite, FeS_2) and the bright (galena, PbS) phases when switching from low to high emission angles (θ_{out}). To our knowledge, this inversion has never been reported in the literature using this set-up. Inversion of Z-contrast was reported recently by Aoyama and co-workers [34], but only at low beam energy and with a $0°$ specimen tilt set-up and a solid-state BSE detector. This type of inversion might be mostly due to some surface contamination. However, in our experiment, high beam energy was used, thus greatly reducing the effect of surface contamination, even with a high-tilt set-up. Flat milling was applied to the specimen to reduce the contamination artefacts due to surface contaminants as much as possible.

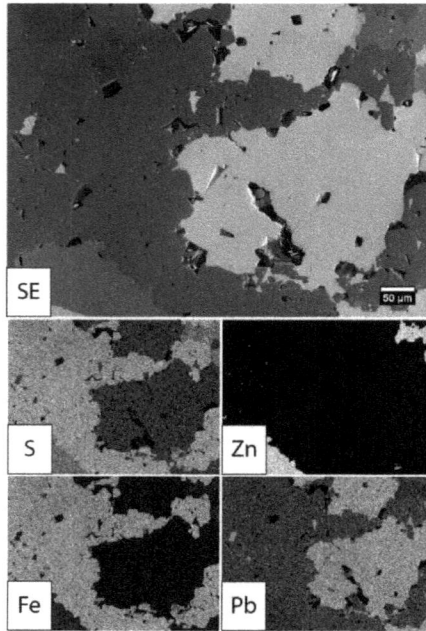

Figure 4. EDS net intensity maps with the corresponding SE image from the galena-rich slag obtained with an accelerating voltage of 5 kV showing that the slag was essentially constituted of iron sulfide, zinc sulfide and lead sulfide. The scalebar displayed on the SE image applies to the corresponding X-ray images. SINGLE COLUMN.

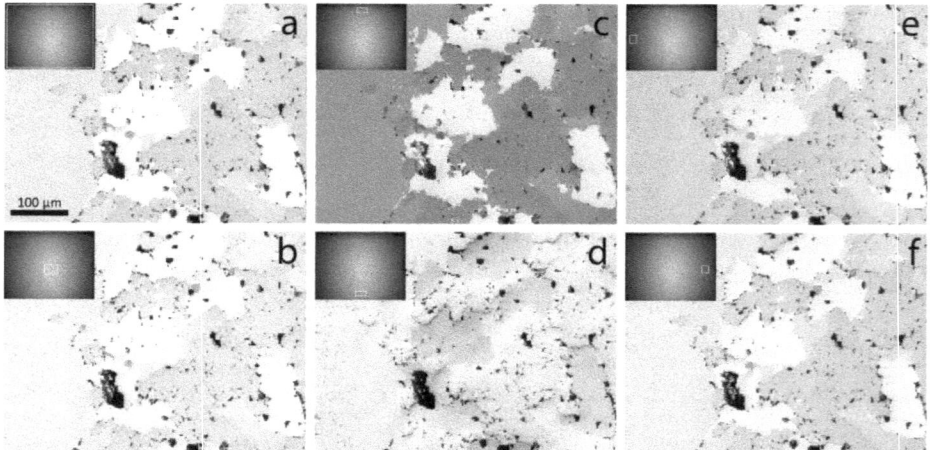

Figure 5. Reconstructed images from an EBSD scan at $E_0 = 20$ kV acquired from a mineral slag sample with the pixel array areas on the reference image (inset) as highlighted by the white squares. (**a**) Full reference image area; (**b**) small area at the center; (**c**) small area from top-center; (**d**) small area from bottom-center; (**e**) small area from left-center and (**f**) small area from right-center. The scalebar displayed on the image (**a**) applies to all the other images of the figure. DOUBLE COLUMN.

When comparing images obtained with pixel arrays from the center left (Figure 5e) and right (Figure 5f) areas of the reference image, no significant change of contrast could be observed; only the illumination intensity due to a slight shift between the two positions regarding the maximum intensity position. Also, one may notice that the contrast from these two images was similar to that of the image obtained from the central position (Figure 5b). Based on this finding, large azimuthal collection angles were used to generate the images shown in Figure 6, while the angle along the y-axis of the reference image was kept similar to that in Figure 5b–d. This allowed us to reduce statistical noise and increase the visibility on the compositional contrast inversion, in addition to compensating for the illumination variation along the camera x-axis observed in Figure 5e–f.

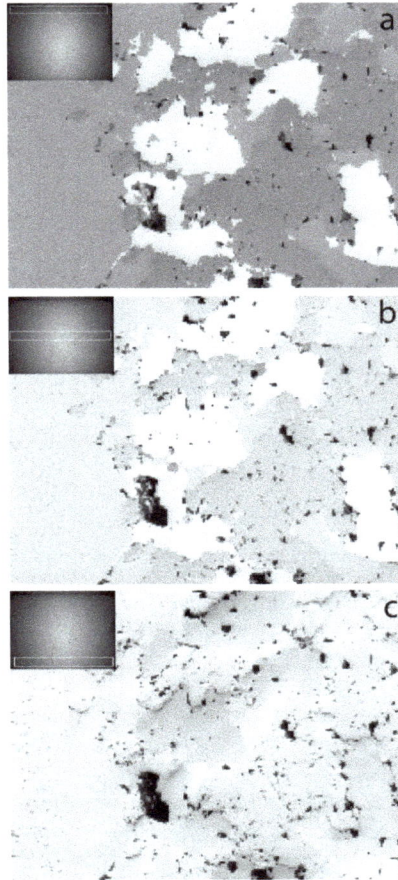

Figure 6. Reconstructed images from the same EBSD scan as in Figure 5. The white rectangles in the reference image shown in the inset define the pixel arrays integrated to obtain each image. (**a**) Top; (**b**) center and (**c**) bottom region of the EBSD camera. The scalebar displayed on the image in Figure 5a applies to all the reconstructed images of this figure.

3.3. Magnetic Domain Imaging

There are several ways to image the magnetic domain structures of ferro-magnetic materials like iron, cobalt, nickel, and their alloys, as well as manganese alloys. A review of these techniques is available in [2]. While contrast of type I is observed due to stray fields generated by the domains above the specimen surface [35,36], the closed magnetic domains from a polished surface can be imaged with magnetic contrast of type II [37,38]. This contrast arises because of the interaction of the internal magnetic field of the specimen with the beam electrons diffusing through the specimen. In fact, maximum contrast between domains of opposite magnetization is obtained when the magnetic field vector, $\overline{B}$, is parallel to the tilt axis of the microscope and the specimen tilted to 50–70° [26,27]. The Lorentz force applied to each primary electron inside the specimen deviates its path inward (Figure 7a) or outward (Figure 7b) the exit surface, depending on the direction of $\overline{B}$, compared to a situation without magnetic force. This mechanism modulates the BSE yield captured by the BSE detector for each opposite domain.

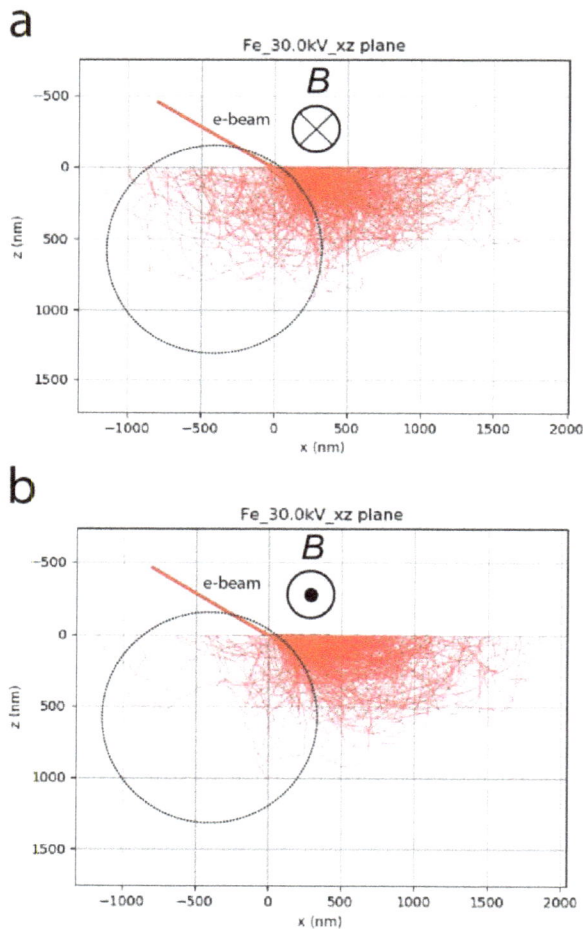

Figure 7. Monte Carlo simulations of electron trajectories in pure iron with 2000 trajectories displayed with the internal magnetic field vector $\overline{B}$ parallel to the tilt axis pointing in the direction into (**a**) and out of (**b**) the plane of the figure. Tilt angle was 70° as in a conventional EBSD set-up. The strength of the magnetic induction was deliberately set to 100 T to enhance the magnetic deflection for ease of visualization. The deflection effect of the internal magnetic field of the specimen is clearly seen in the region indicated by the black dotted circles.

Recently, there has been a renewed interest in magnetic domain imaging using contrast of type II, because the optimum specimen tilt and beam energy conditions are similar to those necessary for analyzing crystal orientation by conventional EBSD. The combination of these two techniques makes it possible to relate the observed specific magnetic domain structures with the angles between the specimen surface and the closest easy axis of each grain [28]. This was accomplished without specialized equipment; only an EBSD camera equipped with a set of two solid-state diodes (left side of Figure 8a), known as the forward scatter detector, were necessary. However, although Monte Carlo modeling did reveal the overall mechanisms of production of this contrast more than 40 years ago, optimization of the detection set-up has still not been reported. To this end, we applied the method described above to acquire a set of EBSPs with a single EBSD scan over five magnetic domains from a NOES sample previously located using the FSD attached to the EBSD camera screen (right side of Figure 8a). The EBSPs were recorded with an image resolution of 128 pixels × 96 pixels and no background correction applied, i.e., only raw patterns were recorded. The pixel dwell time and probe current were carefully optimized to avoid saturation of the phosphorescent screen.

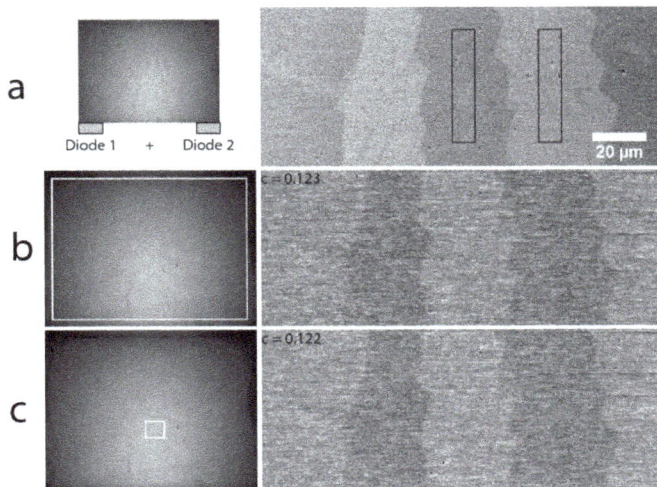

Figure 8. Magnetic domain images obtained from a non-oriented Fe-2.78%-Si electrical steel (right) and collection areas used to produce the images (left) with (**a**) the forward scatter detector (diode 1 + diode 2) attached to the EBSD camera screen; (**b**) the summed intensities over the full area of the EBSD camera screen and (**c**) the summed intensities from a small rectangle located around the maximum illumination area. The black rectangles in (**a**) define the areas used to calculate the contrast *C* displayed in (**b**–**c**). The scale bar displayed on the FSD image in (**a**) applies to the images in (**b**–**c**).

The images resulting from using the whole screen area (Figure 8b) and only a small squared area around the maximum of intensity (Figure 8c) are displayed in Figure 8. Note that the Kikuchi bands are clearly visible on the EBSPs even without a background subtraction procedure. These images correspond to the same specimen area as previously imaged with the FSD and given in Figure 8a. Obviously, both images provide a similar contrast *C* as indicated on each image, i.e., 0.123 (Figure 8b) and 0.122 (Figure 8c). The contrast was calculated with the integrated intensities from the two areas indicated in Figure 8a by the black rectangles. Note that the contrasts reported here are of the same order as those reported previously in the literature [27].

As in the case of compositional contrast studied in Section 3.2, small squared pixel arrays were integrated at different locations of the captured EBSP image. The resulting reconstructed images are shown in Figure 9, with the domain images on the right side and the reference EBSP image on the left side with the pixel array defined by the white squares. Contrast values between the domains of opposite magnetization obtained from Equation (1) are given for each domain image. The areas integrated to compute the contrast were identical to those indicated in Figure 8a. Clearly, the location of the collected area on the EBSP image had a dramatic effect on the final domain images. As stated in Section 3.2 regarding the compositional contrast, the polar angle (along the y-axis of the EBSP image) modulates the level of contrast on the domain images. When the top array was collected, the contrast dropped to 0.047 compared to 0.122 obtained with the array centered on the maximum illumination area. This value indicates that the domains become barely visible when high emission angles are collected. In contrast, the image obtained with the collection array located at the bottom of the reference image, i.e., below the maximum of intensity, showed a higher contrast value at 0.166, which expresses an approximate relative increase of contrast of 36%. This indicates that low polar angles should be used to collect the BSE signal in order to maximize magnetic domain contrast of BSE images.

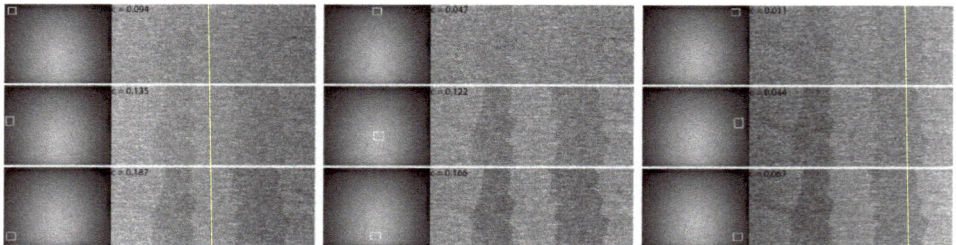

Figure 9. Magnetic domain images (right) obtained from the same EBSD scan as Figure 8. The reference image is displayed on the left of each domain image with the white square defining the collection area used for the image reconstruction. The scalebar displayed on the FSD image in Figure 8a applies to all the images of the figure.

This finding was clearly reproduced when the same pixel array locations were used but shifted to the left or right sides of the reference image (along the x-axis of the image). It must be noted that, due to the extreme locations of the boxes on the left and right positions, non-uniform illumination was observed between the left and right sides of the reconstructed images. Thus, because the contrast was calculated between the central domain and that sitting on its right (Figure 8a), the contrasts values calculated for the images on the right side of Figure 9 were lower than those at the right side of the figure. Although the contrasts values were slightly different for the reason described above, the effect of changing the position of the pixel array along the x-axis of the reference image had a negligible visual effect on the observed contrast.

3.4. EBSD-Dark-Field Applied to Transmission Electron Forward Scattering Diffraction

Transmission EBSD (t-EBSD, t-EFSD, TKD) is an emerging technique and was first reported by Keller and co-workers [39]. It uses a commercial EBSD camera to collect the transmitted diffracted electrons to produce transmission Kikuchi patterns [39–41]. This technique is capable of using the full EBSD data processing capabilities on thin electron transparent specimens, in addition to dramatically improving the spatial resolution of orientation mapping to values directly comparable with those obtained with a transmission electron microscope (TEM) [21,22,42,43]. However, the data collected using this technique has not been utilized to its real potential, since most of the work reported so far was concentrated on acquiring orientation or phase maps.

In this section, we report on some preliminary work regarding the use of the EBSD-DF technique to extract useful images from a t-EBSD dataset. A bright-field image recorded using the transmission mode of the SEM, as well as the corresponding band contrast map resulting from a t-EBSD scan on the same area, are shown in Figure 10a,b, respectively. The sample was a thin foil of an AA2099 Al-Li-Cu alloy of approximately 100 nm thickness. The large black precipitates (few hundreds of nm) are Al_2Cu, and the very tiny ones are T1 precipitates (50–80 nm), namely Al_2CuLi. Note that, possibly due to residual stresses inside some grains that might have been released during the thinning process, the grain in Figure 10 appeared bent, with varying contrast along the grain. Thus, because the T1 precipitates generate a strain field around the precipitates, this allows us to visualize them if a specific diffraction condition is used. The acquisition software was used to produce the band contrast map, and a white cross was marked at the location where the reference pattern displayed in Figure 11 was taken.

One important characteristic of BF imaging is that the diffraction contrast is predominant for crystalline specimens, which complicates the identification of precipitates by relying only on the Z-contrast. However, when the EBSD-DF technique is applied, different conditions can be tested to obtain many dark-field (DF) images with different contrasts from the same area, allowing us to optimize and better understand the microstructure under investigation. DF images generated from the same set of EBSD data as used in Figure 10 are shown in Figure 11a–i. For each image, the t-EBSP image is given on the top, while the generated DF image is at the bottom. Note that the t-EBSP image was cropped to keep only useful diffraction information, i.e., the region of the camera providing Kikuchi contrast.

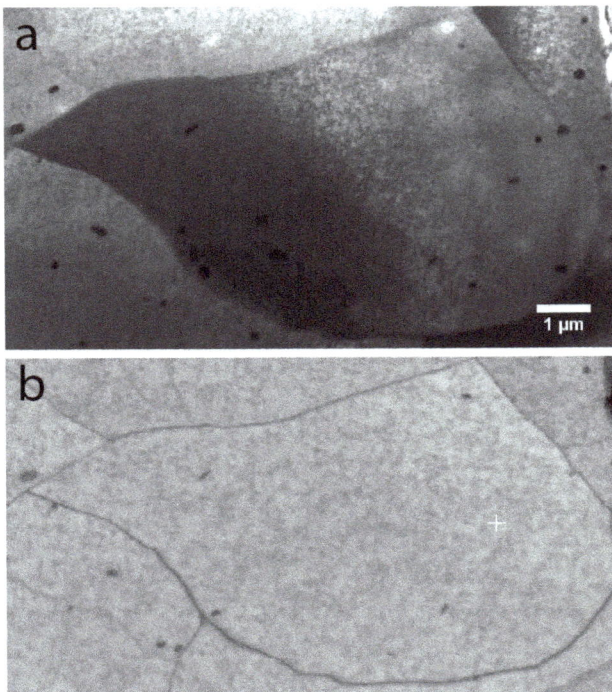

Figure 10. (**a**) Bright-field image of an AA2099 Al-Li-Cu alloy using the transmission mode of the SEM; (**b**) Band contrast (BC) image corresponding to the same area by transmission EBSD. The white cross locates the position of the reference EBSP used for pixel array selection for the reconstruction of the EBSD-DF images. The scalebar displayed on the STEM image in (**a**) applies to the BC image in (**b**).

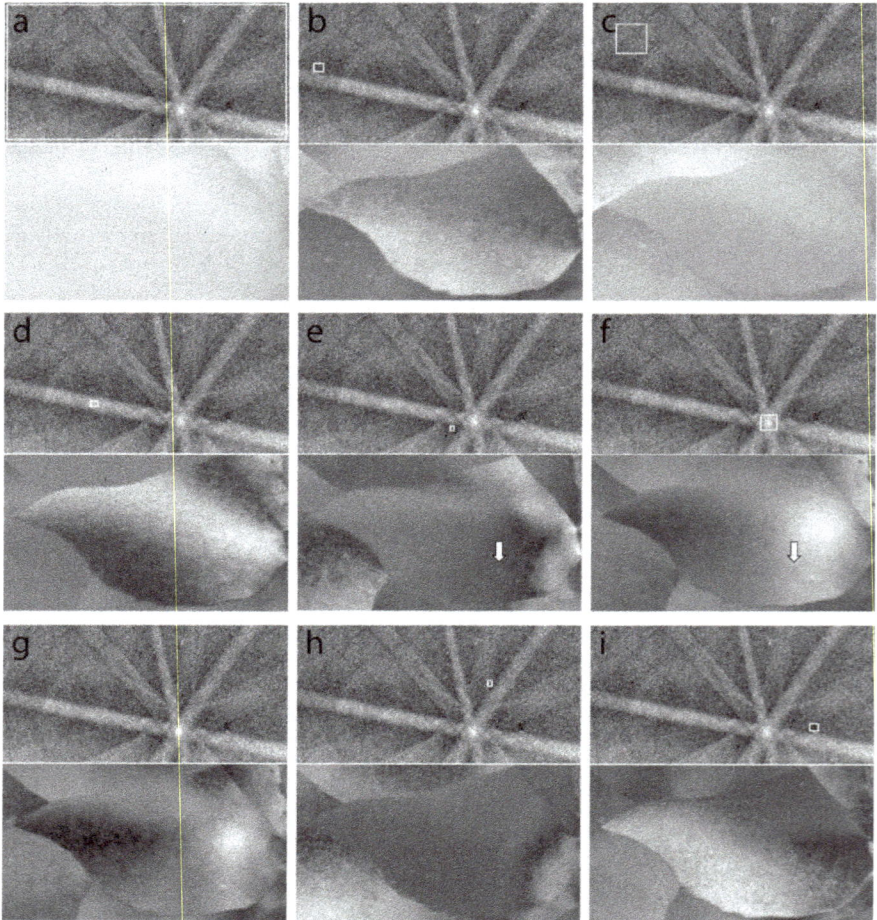

Figure 11. (a–i) EBSD-DF images obtained after collecting the intensities from the pixel arrays (virtual beam) defined by the white squares in the reference Kikuchi pattern (t-EBSP). The top and bottom images correspond to the reference pattern and the EBSD-DF image. The white arrows indicate the precipitates affected by the diffraction condition used (see text for more information). Note that the t-EBSPs were background subtracted to enhance diffraction contrast. The scalebar displayed on the STEM image in Figure 9a applies to all reconstructed images this figure.

When the full t-EBSP area were summed to generate the image in Figure 11a, the contributions from many diffraction conditions were averaged, and the final contrast was very weak, impeding visualization of the true microstructure inside the grain. Similarly, the contrast was very weak when the pixel array was chosen in a region of the t-EBSP with poor Kikuchi band contrast (Figure 11c). On the other hand, if a small sized pixel array was selected in specific areas of the reference pattern, the achievable contrast depended on the specific location chosen. If one looks closely, the intensity at the area where the reference pattern was selected on the reconstructed image (white cross in see Figure 10b) matches the intensity inside of the integrated pixel array in the reference pattern image. For example, when the virtual beam (pixel array) was selected inside a low index zone axis, the zone axis image was observed in the reconstructed image in the same manner as the bend contours in

TEM [44] due to the foil bending. This holds also when the virtual beam is positioned inside a Kikuchi band (Figure 11d), where the image of the selected band sits at the location of the reference image. In Figure 11e, the virtual beam is located just in a corner of the zone axis, and consequently, the image of the zone axis appears shifted to the right of the reference pattern position. Note that the contrast is observed here was driven by the crystal rotation due to the foil bending which moved the Kikuchi pattern captured by the EBSD camera around the virtual beam position for every pixel of the grain. It must also be noticed that some precipitates visible in the bright-field image were not observed in any of the reconstructed images in Figure 11. This is due to the depth origin of the diffracted forward scattered electrons captured by the EBSD screen. In fact, the depth resolution of diffracted beams is limited by the inelastic scattering which takes place after the Bragg diffraction events, and thus, only the forward scattered electrons diffracted close to the exit surface are allowed to exit with coherency conservation. This effect must be kept in mind when our method is combined to t-EBSD.

In addition to the bend contour effect, the diffraction conditions used for producing the DF images have a significant influence on the visibility of precipitates in the matrix. Contrast inversion may be observed as in Figure 11, where the two large precipitates at the bottom left of the grain appear either bright on a dark matrix (Figure 11e, see arrow) or dark on a bright matrix (Figure 11f, see arrow). Also, the fine T1 precipitates—that are visible because of the lattice distortions they generate due to the foil bending—were affected by the position of the virtual beam. When a zone axis or the central area of a Kikuchi band are selected, the precipitates are nearly extinguished (Figure 11d,f,g) while they clearly appeared bright over a dark background when the virtual beam was moved to the dark side of the Kikuchi lines (Figure 11e,h,i). This clearly demonstrates the usefulness of the post-processing technique described in this work to better qualify the microstructure of a complex alloy like the one characterized here.

4. Discussion

Its large collection angle, either polar or azimuthal, combined to the pixelization of the collection area make the commercial EBSD camera an ideal device for spatially filtering the backscattered electrons signal emitted from a highly tilted specimen. In addition, the pattern storing and binning facilities of this type of detector makes it efficient in monitoring contrasts over the range of angles collected by the camera, where many images from the same scan can be produced from the exact same specimen area in only one scan. This greatly limits the carbon contamination problems that are often encountered when several scans must be carried out from the exact same region for comparison purposes. With the best cameras currently available, acquisition rates of more than 1000 pattern/s are common, and the latest systems with CMOS technology are capable of acquiring a 1000×1000 image in less than 10 min [45]. This means that the time required to record an image becomes comparable to that necessary to record a BSE image with high signal-to-noise ratio. Work in progress on direct detection cameras for EBSD must also be considered, as it might bring a new dimension through energy filtering possibilities provided by this type of detection [46,47].

The experimental results regarding compositional and topographical contrast as a function of the polar collection angle are consistent with previously reported work [9,15,48]. Hence, topographical contrast is enhanced at low θ_{out} emission angles. This is due to the increased shadowing on the emitted BSEs at grazing exit angles which are more strongly absorbed than their higher angles counterparts. In addition, these electrons have a limited number of elastic collision in comparison to those emitted at higher angles. In fact, Z-contrast is obtained through the modulation of the emitted intensity depending on Z. At high Z, primary electrons have larger angle elastic collisions, allowing them to reach the surface and be emitted. The primary electrons travelling in a low Z material suffer less large angle elastic collisions, and are thus more absorbed in the matter. The number of emitted BSEs is then lower than that emitted from the high Z material. In high tilt conditions, the number of BSE emitted at small angles (θ_{out}) increases due to the reduced path to the surface. However, the BSEs emitted at higher angles suffer more elastic collisions, resulting in larger total deviation (longer path to the surface), and thus, a higher probability of being emitted at a high angle (θ_{out}). In consequence, the Z-contrast is stronger at high emission angles. The interesting point that comes out of our experimental results is that compositional contrast inversion was observed at very low angles (30°–40°), even if this contrast was very weak and mostly dominated by topographical contrast. This, to our knowledge, has never been reported, and further work needs to be carried on to confirm this finding. One could suspect channeling contrast as the origin of the contrast inversion, because channeling contrast is stronger at low emission angle. However, the contrast inversion was observed for all grains of the same phase everywhere in the map, and no channeling effects were evidenced. Finally, it must be noted that the azimuthal angle of emission, i.e., the angle around the beam axis, has no effect on the compositional contrast observed with the EBSD detector. This is expected, as the interaction mechanism driving the emission of the BSEs is the same for mirror positions on both sides of the beam impact point as long as the beam is normal to the tilt axis, and no internal magnetic field is present in the specimen.

Magnetic domain imaging was studied in detail more than 40 years ago. The optimum SEM parameters to observe magnetic domain contrast of type II were found to be preferentially high beam accelerating voltage, high specimen tilt, and with the internal magnetic field vector $\bar{B}$ parallel to the tilt axis [27,37,49]. In addition, the contrast between domains is enhanced if energy filtration is applied to the BSE signal, since most of the contrast is carried out by the low-loss BSEs [27,50,51]. However, the effect of the BSE detector collection angles (polar and azimuthal) was not reported, and the preliminary results reported in this work gives new information regarding the understanding of type II magnetic contrast. Hence, the highest contrast was observed at low polar collection angles (θ_{out}), meaning a lower angle than that corresponding to the maximum BSE intensity. This suggests that the contrast is maximal when slightly deviated primary electrons are collected, since the bottom area used in Figure 9 (bottom center image) corresponds to collection angles of 30–40° with respect to the specimen surface. According to Wells [50], the low-loss BSEs, mainly responsible for type II contrast, are primary electrons suffering only a single large angle elastic collision, the probability of which decreases with emission angle. The existence of an optimum collection angle for magnetic domain imaging of ferromagnetic materials needs to be confirmed with further experiments and compared with advanced Monte Carlo simulations. The impact of the azimuthal angle also needs to be understood with regard to the internal magnetic field direction, although our results tend to show that it has no effect on our specific example. Thus, by using our approach, the optimum collection parameters for high type II contrast can be determined for any EBSD/FSD system, and this will help in positioning the FSD diodes to get the best contrast. It has to be noted, furthermore, that type II magnetic contrast was reported with 0° tilt angles with a below-the-lens solid state BSE detector on Ni-Mn-Ga alloys, which was not predicted by the simulations performed by Newbury [52,53]. More attention needs to be given to this specific case, and further research on this subject is under way.

The last example that was investigated in Section 3.4 showed the importance of the selected diffraction conditions on the reference EBSP on the resulting images of the microstructure and precipitates. If the summed intensity at the pixel array area of the reference EBSP brings a bright background (BF mode), it may vanish the contrast from lighter precipitates that appear brighter than the matrix if material contrast is considered. On the other hand, if the selected area brings a dark matrix background, the contrast from the same precipitates will be enhanced. The mechanism holds with reversed precipitates contrast. This was demonstrated here with the Al_2Cu precipitates with higher mean atomic number compared to the aluminum matrix. This effect has already been reported with the same alloy in bulk form, where BSE imaging at low accelerating voltage was applied with high channeling conditions [54]. In this case, the δ' Al_3Li precipitates were lighter than the Al matrix, but the contrast was found to vary depending on the channeling conditions applied to the Al grain. The same effect was noticed with iron carbides in steel (work not reported). Thus, controlling the background backscattered or transmitted electrons emission in a post-process fashion allows one to monitor precipitates/matrix contrasts, in order to bring a specific area into contrast depending on the area chosen to extract the reference EBSP. This is convenient, especially for deformed materials, where the changing lattice orientation across a grain is detrimental to the extraction of a satisfactory precipitate's contrast from a single image over the whole grain surface. The availability of a large number of images with different diffraction conditions is capable of resolving the microstructure more finely with only a single scan when our method is applied. This confirms, as already reported for bulk specimens [17], that true DF images in transmission mode can be obtained via this technique where the contrast can be directly related to the diffraction condition used.

5. Conclusions

In this work, an EBSD camera was turned into a BSE detector with a large polar collection angle ranging from around 30° to 100°. Thanks to the storing capability of this equipment, each EBSP image was recorded and stored in a datacube type data set using a dedicated Python program based on the HDF5 export library. Due to the pixelization of the detector frame, the detector captures the BSE angular distribution of the signal emitted from specimens tilted up to 70° with respect to the electron beam, and allows to spatially filter the collected BSE intensity. Thus, for any pixel array selected in the reference EBSP image, a reconstructed image can be produced rapidly by summing the contribution of all the pixels of the array, thanks to the HDF5 datacube format used to store the EBSPs. In this paper, we demonstrated that this large area detector can be successfully used to optimize backscattered or transmitted electrons contrasts of different types; our findings can be summarized as follows:

1. In conventional mode, i.e., with highly tilted bulk specimens, our results confirm the previously reported findings regarding the contrast obtained versus the polar collection angle of the camera. High emission angles (θ_{out}) with respect to the specimen surface are prone to bring compositional contrast into the reconstructed image, while small emission angles carry the topographic contrast component. However, compositional contrast inversion was found when very small collection angles were collected.
2. Magnetic domain contrast imaging at elevated tilt angles was optimized by comparing images obtained at different polar emission angles. The highest contrast was obtained with the BSEs emitted at the lowest angles with respect to the specimen surface. Further investigations need to be carried out to confirm, and maybe improve, the contrast if possible.
3. When the reference image captured by the EBSD camera arises from a crystalline material, the reconstructed images carry the diffraction information related to the specific reflection excited via the virtual beam represented by the pixel array chosen in the reference EBSP. The resulting images thus mimics electron channeling contrast, and allows us to visualize deformation in materials in a new way thanks to the many multiple images that can be generated with this technique from a single scan.
4. The diffraction contrast was applied to the transmission mode (t-EBSD) and was capable of generating real transmission dark-field images where the contrast relates to the reflection selected in the reference pattern. The impact on the visibility of fine precipitates inside the matrix was demonstrated, and again, the importance of selecting many different reflections from a single scan was shown to be efficient in characterizing the fine microstructure of a material.

The main limitation of our method is the acquisition time of the EBSD dataset necessary to obtain these images after post-processing; this is mainly due to limitations of the EBSD system hardware. However, the new generation of EBSD cameras have fast mapping capabilities, and the development of direct electron detection cameras for EBSD will bring this method to the level of routine application, especially if energy filtration is coupled with the selection of the collection angles.

Finally, thanks to these preliminary and encouraging results presented in this work, future work needs to be carried out to make the method more robust. A study of the intake of big data analysis to reduce the amount of useful data should be carried out. A reflection on different types of shape, and how they could be combined to generate new contrasts, must also be undertaken. The possibility of choosing a uniform emission angle over the EBSD screen is another potential development to allow us to better master contrast.

Author Contributions: N.B, H.D. and R.G.; Methodology, N.B, H.D. and R.G.; Software, N.B, H.D. and R.G.; Validation, N.B, H.D. and R.G.; Formal Analysis, N.B, H.D. and R.G.; Investigation, N.B, H.D. and R.G.; Resources, N.B, H.D. and R.G.; Data Curation, N.B, H.D. and R.G.; Writing—Original Draft Preparation, N.B, H.D. and R.G.; Writing—Review \& Editing, N.B, H.D. and R.G.; Visualization, N.B, H.D. and R.G.; Supervision, R.G.; Project Administration, N.B, H.D. and R.G.; Funding Acquisition, No funding.

Funding: This research received no external funding.

Conflicts of Interest: The authors declare no conflict of interest.

References

1. Bell:, D.C.; Erdman, N. *Low Voltage Electron Microscopy: Principles and Applications*; John Wiley & Sons: Hoboken, NJ, USA, 2012.
2. Brodusch, N.; Demers, H.; Gauvin, R. *Field Emission Scanning Electron Microscopy: New Perspectives for Materials Characterization*; Springer: Singapore, 2017; p. 137.
3. Coates, D.G. Kikuchi-like reflection patterns obtained with the scanning electron microscope. *Philos. Mag.* **1967**, *16*, 1179–1184. [CrossRef]
4. Venables, J.A.; Harland, C.J. Electron back-scattering patterns—A new technique for obtaining crystallographic information in the scanning electron microscope. *Philos. Mag.* **1973**, *27*, 1193–1200. [CrossRef]
5. Schwarzer, R.A.; Field, D.P.; Adams, B.L.; Kumar, M.; Schwartz, A.J. Present state of electron backscatter diffraction and prospective developments. In *Electron Backscatter Diffraction in Materials Science*; Schwartz, A.J., Kumar, M., Adams, B.L., Field, D.P., Eds.; Springer: Berlin, Germany, 2009; pp. 1–20.
6. Schwartz, A.J.; Kumar, M.; Adams, B.L.; Field, D.P. *Electron Backscatter Diffraction in Materials Science*; Springer: Berlin, Germany, 2009.
7. Zaefferer, S. On the formation mechanisms, spatial resolution and intensity of backscatter kikuchi patterns. *Ultramicroscopy* **2007**, *107*, 254–266. [CrossRef] [PubMed]
8. Steinmetz, D.R.; Zaefferer, S. Towards ultrahigh resolution EBSD by low accelerating voltage. *Mater. Sci. Technol.* **2010**, *26*, 640–645. [CrossRef]
9. Prior, D.J.; Trimby, P.; Weber, U.; Dingley, D.J. Orientation contrast imaging of microstructures in rocks using forescatter detectors in the scanning electron microscope. *Miner. Mag.* **1996**, *60*, 859–869. [CrossRef]
10. Payton, E.J.; Nolze, G. The backscatter electron signal as an additional tool for phase segmentation in electron backscatter diffraction. *Microsc. Microanal.* **2013**, *19*, 929–941. [CrossRef] [PubMed]
11. Wells, O.C.; Gignac, L.M.; Murray, C.E.; Frye, A.; Bruley, J. Use of backscattered electron detector arrays for forming backscattered electron images in the scanning electron microscope. *Scanning* **2006**, *28*, 27–31. [CrossRef] [PubMed]
12. Schwarzer, R.A.; Hjelen, J. Backscattered electron imaging with an EBSD detector. *Microsc. Today* **2015**, *23*, 12–17. [CrossRef]
13. Schwarzer, R.A.; Sukkau, J. Electron back scattered diffraction: Current state, prospects and comparison with X-ray diffraction texture measurement. *Banaras Metall.* **2013**, *18*, 1–11.
14. Nowell, M.M.; Wright, S.I.; Rampton, T.; de Kloe, R. A new microstructural imaging approach through EBSD pattern region of interest analysis. *Microsc. Microanal.* **2014**, *20*, 1116–1117. [CrossRef]
15. Wright, S.I.; Nowell, M.M.; de Kloe, R.; Camus, P.; Rampton, T. Electron imaging with an EBSD detector. *Ultramicroscopy* **2015**, *148*, 132–145. [CrossRef] [PubMed]
16. Brodusch, N.; Demers, H.; Gauvin, R. Dark-field imaging based on post-processing of electron backscatter diffraction patterns in a scanning electron microscope. *Microsc. Microanal.* **2015**, *21*, 2031–2032. [CrossRef]
17. Brodusch, N.; Demers, H.; Gauvin, R. Dark-field imaging based on post-processed electron backscatter diffraction patterns of bulk crystalline materials in a scanning electron microscope. *Ultramicroscopy* **2015**, *148*, 123–131. [CrossRef] [PubMed]
18. Chapman, M.; Callahan, P.; Graef, M. EBSD surface topography determination in a martensitic Au-Cu-Zn alloy. *Microsc. Microanal.* **2015**, *21*, 2215–2216. [CrossRef]

19. Chapman, M.; Callahan, P.G.; De Graef, M. Determination of sample surface topography using electron back-scatter diffraction patterns. *Scr. Mater.* **2016**, *120*, 23–26. [CrossRef]
20. Dorri, M.; Turgeon, S.; Brodusch, N.; Cloutier, M.; Chevallier, P.; Gauvin, R.; Mantovani, D. Characterization of amorphous oxide nano-thick layers on 316l stainless steel by electron channeling contrast imaging and electron backscatter diffraction. *Microsc. Microanal.* **2016**, *22*, 997–1006. [CrossRef] [PubMed]
21. Brodusch, N.; Demers, H.; Gauvin, R. Nanometres-resolution kikuchi patterns from materials science specimens with transmission electron forward scatter diffraction in the scanning electron microscope. *J. Microsc.* **2013**, *250*, 1–14. [CrossRef] [PubMed]
22. Brodusch, N.; Demers, H.; Trudeau, M.; Gauvin, R. Acquisition parameters optimization of a transmission electron forward scatter diffraction system in a cold-field emission scanning electron microscope for nanomaterials characterization. *Scanning* **2013**, *35*, 375–386. [CrossRef] [PubMed]
23. Gauvin, R.; Lifshin, E.; Demers, H.; Horny, P.; Campbell, H. Win X-ray: A new Monte Carlo program that computes X-ray spectra obtained with a scanning electron microscope. *Microsc. Microanal.* **2006**, *12*, 49–64. [CrossRef] [PubMed]
24. Joy, D.C.; Luo, S. An empirical stopping power relationship for low-energy electrons. *Scanning* **1989**, *11*, 176–180. [CrossRef]
25. Heinrich, K.F.; Newbury, D.E.; Yakowitz, H. *Use of Monte Carlo Calculations in Electron Probe Microanalysis and Scanning Electron Microscopy: Proceedings of a Workshop Held at the National Bureau of Standards, Gaithersburg, Maryland, October 1–3, 1975*; US Dept. of Commerce, National Bureau of Standards: Washington, DC, USA; US Govt. Print. Off.: Washington, DC, USA, 1976.
26. Newbury, D.E.; Yakowitz, H.; Myklebust, R.L. Monte carlo calculations of magnetic contrast from cubic materials in the scanning electron microscope. *Appl. Phys. Lett.* **1973**, *23*, 488–490. [CrossRef]
27. Newbury, D.E.; Yakowitz, H.; Myklebust, R.L. A study of type ii magnetic domain contrast in the SEM by Monte Carlo electron trajectory simulation. In *Use of Monte Carlo Calculations in Electron Probe Microanalysis and Scanning Electron Microscopy: Proceedings of a Workshop Held at the National Bureau of Standards, Gaithersburg, Maryland, October 1–3, 1975*; US Dept. of Commerce, National Bureau of Standards: Washington, DC, USA; US Govt. Print. Off.: Washington, DC, USA, 1976.
28. Gallaugher, M.; Brodusch, N.; Gauvin, R.; Chromik, R.R. Magnetic domain structure and crystallographic orientation of electrical steels revealed by a forescatter detector and electron backscatter diffraction. *Ultramicroscopy* **2014**, *142*, 40–49. [CrossRef] [PubMed]
29. Reimer, L. Scanning electron microscopy: Physics of image formation and microanalysis. *Meas. Sci. Technol.* **1998**, *11*, 1826. [CrossRef]
30. Wells, O.C. Comparison of different models for the generation of electron backscattering patterns in the scanning electron microscope. *Scanning* **1999**, *21*, 368–371. [CrossRef]
31. Kaboli, S.; Demers, H.; Brodusch, N.; Gauvin, R. Rotation contour contrast reconstruction using electron backscatter diffraction in a scanning electron microscope. *J. Appl. Crystallogr.* **2015**, *48*, 776–785. [CrossRef]
32. Holt, D.B.; Muir, M.D.; Grant, P.R.; Boswarva, I.M. *Quantitative Scanning Electron Microscopy*; Academic Press: London, UK; New York, NY, USA; San Francisco, CA, USA, 1974.
33. Aoyama, T.; Nagoshi, M.; Nagano, H.; Sato, K.; Tachibana, S. Selective backscattered electron imaging of material and channeling contrast in microstructures of scale on low carbon steel controlled by accelerating voltage and take-off angle. *ISIJ Int.* **2011**, *51*, 1487–1491. [CrossRef]
34. Aoyama, T.; Nagoshi, M.; Sato, K. Quantitative analysis of angle-selective backscattering electron image of iron oxide and steel. *Microscopy* **2015**, *64*, 319–325. [CrossRef] [PubMed]
35. Banbury, J.R.; Nixon, W.C. The direct observation of domain structure and magnetic fields in the scanning electron microscope. *J. Sci. Instrum.* **1967**, *44*, 889. [CrossRef]
36. Joy, D.C.; Jakubovics, J.P. Direct observation of magnetic domains by scanning electron microscopy. *Philos. Mag.* **1968**, *17*, 61–69. [CrossRef]
37. Fathers, D.J.; Jakubovics, J.P.; Joy, D.C.; Newbury, D.E.; Yakowitz, H. A new method of observing magnetic domains by scanning electron microscopy I. Theory of the image contrast. *Phys. Status Solidi A* **1973**, *20*, 535–544. [CrossRef]
38. Fathers, D.J.; Jakubovics, J.P.; Joy, D.C.; Newbury, D.E.; Yakowitz, H. A new method of observing magnetic domains by scanning electron microscopy. II. Experimental confirmation of the theory of image contrast. *Phys. Status Solidi A* **1974**, *22*, 609–619. [CrossRef]

39. Geiss, R.H.; Keller, R.R.; Read, D.T. Transmission electron diffraction from nanoparticles, nanowires and thin films in an SEM with conventional EBSD equipment. *Microsc. Microanal.* **2010**, *16*, 1742–1743. [CrossRef]

40. Geiss, R.; Keller, R.; Sitzman, S.; Rice, P. New method of transmission electron diffraction to characterize nanomaterials in the sem. *Microsc. Microanal.* **2011**, *17*, 386–387. [CrossRef]

41. Keller, R.; Geiss, R. Transmission ebsd from 10 nm domains in a scanning electron microscope. *J. Microsc.* **2012**, *245*, 245–251. [CrossRef]

42. Suzuki, S. Features of transmission EBSD and its application. *JOM* **2013**, *65*, 1254–1263. [CrossRef]

43. Trimby, P.W. Orientation mapping of nanostructured materials using transmission kikuchi diffraction in the scanning electron microscope. *Ultramicroscopy* **2012**, *120*, 16–24. [CrossRef] [PubMed]

44. Steeds, J.W.; Tatlock, G.J.; Hampson, J. Real space crystallography. *Nature* **1973**, *241*, 435. [CrossRef]

45. Oxford_Instruments. Rapid Characterization of Steel and Ni. Application Note 2017. Available online: http://symmetry.oxford-instruments.com/ (accessed on 1 March 2018).

46. Vespucci, S.; Winkelmann, A.; Naresh-Kumar, G.; Mingard, K.P.; Maneuski, D.; Edwards, P.R.; Day, A.P.; O'Shea, V.; Trager-Cowan, C. Digital direct electron imaging of energy-filtered electron backscatter diffraction patterns. *Phys. Rev. B* **2015**, *92*, 205301. [CrossRef]

47. Wilkinson, A.J.; Moldovan, G.; Britton, T.B.; Bewick, A.; Clough, R.; Kirkland, A.I. Direct detection of electron backscatter diffraction patterns. *Phys. Rev. Lett.* **2013**, *111*, 065506. [CrossRef] [PubMed]

48. Reimer, L.; Riepenhausen, M.; Schierjott, M. Signal of backscattered electrons at edges and surface steps in dependence on surface tilt and take-off direction. *Scanning* **1986**, *8*, 164–175. [CrossRef]

49. Yamamoto, T.; Nishizawa, H.; Tsuno, K. Magnetic domain contrast in backscattered electron images obtained with a scanning electron microscope. *Philos. Mag.* **1976**, *34*, 311–325. [CrossRef]

50. Wells, O. Calculation of type ii magnetic contrast in the low-loss image in the scanning electron microscope. In *Use of Monte Carlo Calculations in Electron Probe Microanalysis and Scanning Electron Microscopy: Proceedings of a Workshop Held at the National Bureau of Standards, Gaithersburg, Maryland, October 1–3, 1975*; US Dept. of Commerce, National Bureau of Standards: Washington, DC, USA; US Govt. Print. Off.: Washington, DC, USA, 1976.

51. Wells, O.C.; Savoy, R.J. Enhancement of type-2 magnetic contrast in the BSE image in the SEM by a lock-in technique. *Scanning* **1979**, *2*, 255–256. [CrossRef]

52. Ge, Y.; Heczko, O.; Soderberg, O.; Hannula, S.P.; Lindroos, V.K. Investigation of magnetic domains in Ni–Mn–Ga alloys with a scanning electron microscope. *Smart Mater. Struct.* **2005**, *14*, S211. [CrossRef]

53. Ge, Y.; Heczko, O.; Soderberg, O.; Lindroos, V. Various magnetic domain structures in a Ni-Mn-Ga martensite exhibiting magnetic shape memory effect. *J. Appl. Phys.* **2004**, *96*, 2159–2163. [CrossRef]

54. Brodusch, N.; Voisard, F.; Gauvin, R. About the contrast of δ' precipitates in bulk Al-Cu-Li alloys in reflection mode with a field-emission scanning electron microscope at low accelerating voltage. *J. Microsc.* **2017**, *268*, 107–118. [CrossRef] [PubMed]

MDPI
St. Alban-Anlage 66
4052 Basel
Switzerland
Tel. +41 61 683 77 34
Fax +41 61 302 89 18
www.mdpi.com

Journal of Imaging Editorial Office
E-mail: jimaging@mdpi.com
www.mdpi.com/journal/jimaging

www.ingramcontent.com/pod-product-compliance
Lightning Source LLC
Chambersburg PA
CBHW051908210326
41597CB00033B/6074